Solutions Manual to Accompany

CLASSICAL GEOMETRY

Euclidean, Transformational, Inversive, and Projective

I. E. Leonard

J. E. Lewis

A. C. F. Liu

G. W. Tokarsky

Department of Mathematical and Statistical Sciences
University of Alberta
Edmonton, Canada

WILEY

Published by John Wiley & Sons, Inc., Hoboken, New Jersey.
Published simultaneously in Canada.

For general information on our other products and services please contact our Customer Care Department within the United States at (800) 762-2974, outside the United States at (317) 572-3993 or fax (317) 572-4002.

Wiley also publishes its books in a variety of electronic formats. Some content that appears in print, however, may not be available in electronic formats. For more information about Wiley products, visit our web site at www.wiley.com.

Library of Congress Cataloging-in-Publication Data:

Leonard, I. Ed., 1938– author.
Solutions manual to accompany classical geometry : Euclidean, transformational, inversive, and projective / I. E. Leonard, Department of Mathematical and Statistical Sciences, University of Alberta, Edmonton, Canada, J.E. Lewis, Department of Mathematical and Statistical Sciences, University of Alberta, Edmonton, Canada, A.C.F. Liu, Department of Mathematical and Statistical Sciences, University of Alberta, Edmonton, Canada, G.W. Tokarsky, Department of Mathematical and Statistical Sciences, University of Alberta, Edmonton, Canada.
pages cm
ISBN 978-1-118-90352-0 (pbk.)
1. Geometry. I. Lewis, J. E. (James Edward) author. II. Liu, A. C. F. (Andrew Chiang-Fung) author. III. Tokarsky, G. W., author. IV. Title.
QA445.L46 2014
516—dc23 2013042035

10 9 8 7 6 5 4 3 2 1

CONTENTS

PART I EUCLIDEAN GEOMETRY

PART II TRANSFORMATIONAL GEOMETRY

PART III INVERSIVE AND PROJECTIVE GEOMETRIES

PART I

EUCLIDEAN GEOMETRY

CHAPTER 1

CONGRUENCY

1. Prove that the internal and external bisectors of the angles of a triangle are perpendicular.

   Solution. Let BD and BE be the angle bisectors, as shown in the diagram below.

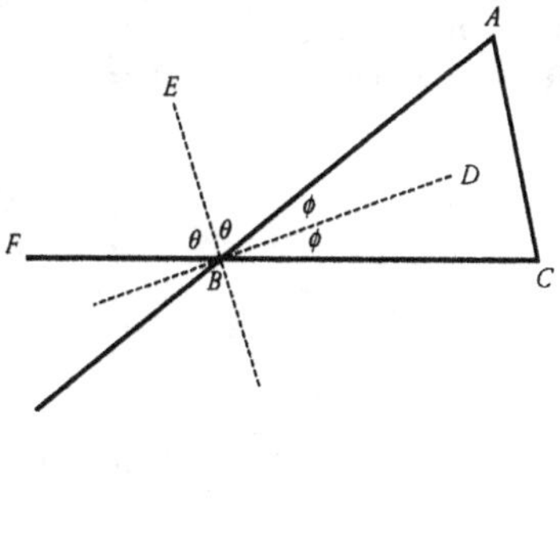

Then

$$\angle EBD = \angle EBA + \angle DBA = \frac{\angle FBA}{2} + \frac{\angle CBA}{2} = \frac{\angle FBA + CBA}{2} = 90.$$

Solutions Manual to Accompany Classical Geometry: Euclidean, Transformational, Inversive, and Projective, First Edition. By I. E. Leonard, J. E. Lewis, A. C. F. Liu, G. W. Tokarsky.

3. Let P be a point inside $\triangle ABC$. Use the Triangle Inequality to prove that $AB + BC > AP + PC$.

Solution. Extend AP to meet BC at D. Using the Triangle Inequality,

$$AB + BD > AD = AP + PD$$

so that

$$AB + BD + DC > AP + PD + DC.$$

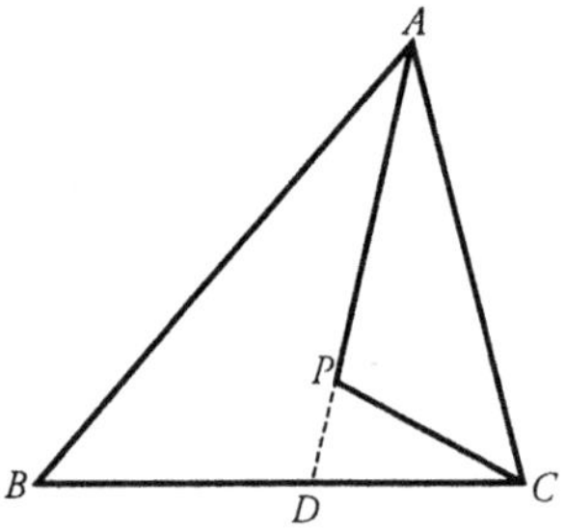

Since

$$BD + DC = BC$$

and

$$PD + DC > PC,$$

we have

$$AB + BC > AP + PC.$$

5. Given the isosceles triangle ABC with $AB = AC$, let D be the foot of the perpendicular from A to BC. Prove that AD bisects $\angle BAC$.

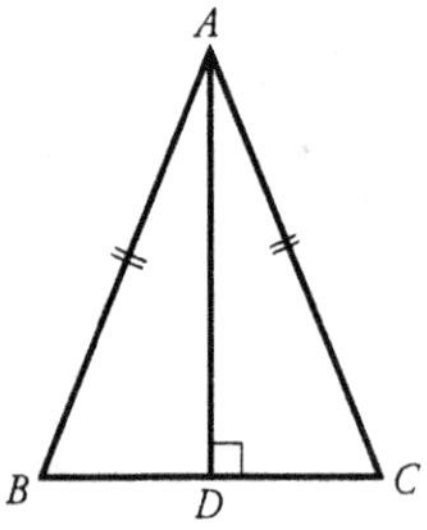

Solution. Referring to the diagram, the two right triangles ADB and ADC have a common side and equal hypotenuses, so they are congruent by **HSR**. Consequently, $\angle BAD \equiv \angle CAD$.

7. D is a point on BC such that AD is the bisector of $\angle A$. Show that

$$\angle ADC = 90 + \frac{\angle B - \angle C}{2}.$$

Solution. Referring to the diagram, $2\theta + \beta + \gamma = 180$, which implies that

$$\theta = 90 - \frac{\beta + \gamma}{2}.$$

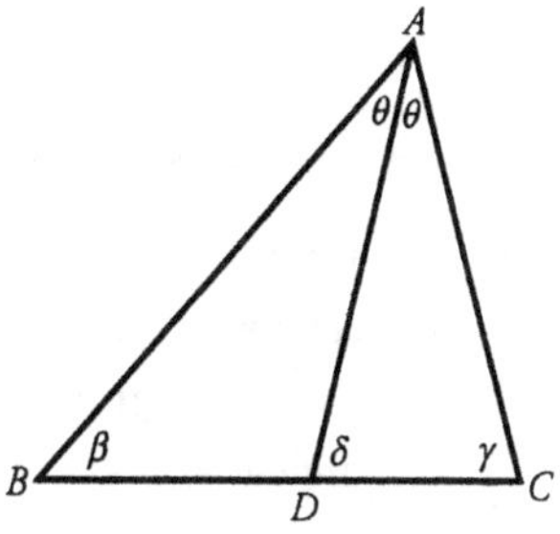

From the Exterior Angle Theorem, we have

$$\delta = \theta + \beta,$$

so that

$$\delta = 90 - \frac{\beta + \gamma}{2} + \beta = 90 + \frac{\beta - \gamma}{2}.$$

9. Construct a right triangle given the hypotenuse and one side.

   Solution. We construct a right triangle ABC given the hypotenuse BC and the length c of side AC.

   Construction.

   (1) Construct the right bisector of BC, yielding M, the midpoint of BC.

   (2) With center M, draw a semicircle with diameter BC.

   (3) With center C and radius equal to c, draw an arc cutting the semicircle at A.

   Then ABC is the desired triangle.

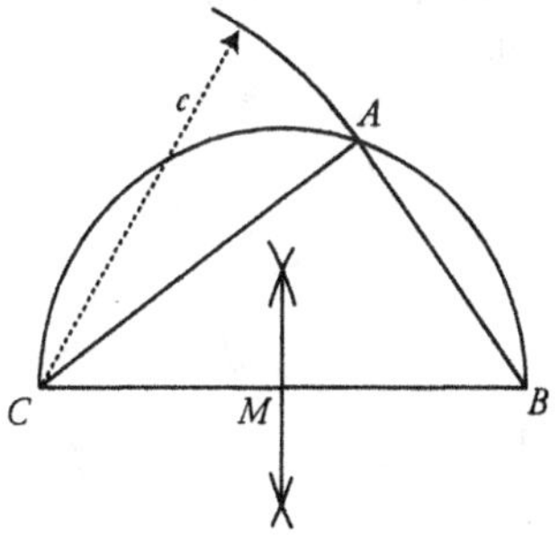

   Justification. $\angle BAC$ is a right angle by Thales' Theorem.

11. Let Q be the foot of the perpendicular from a point P to a line l. Show that Q is the point on l that is closest to P.

    Solution. Let X be any point on l with $X \neq Q$, as in the figure below.

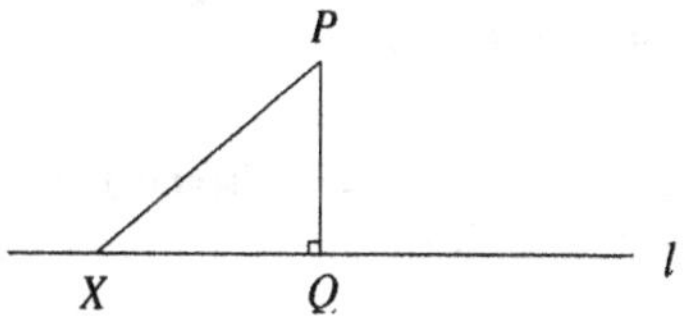

By Pythagoras' Theorem, we have

$$PX^2 = PQ^2 + XQ^2 > PQ^2,$$

and therefore $PX > PQ$.

13. Let $ABCD$ be a simple quadrilateral. Show that $ABCD$ is cyclic if and only if the opposite angles sum to $180°$.

Solution. We will show that the simple quadilateral $ABCD$ can be inscribed in a circle if and only if $\angle A + \angle C = 180$ and $\angle B + \angle D = 180$.

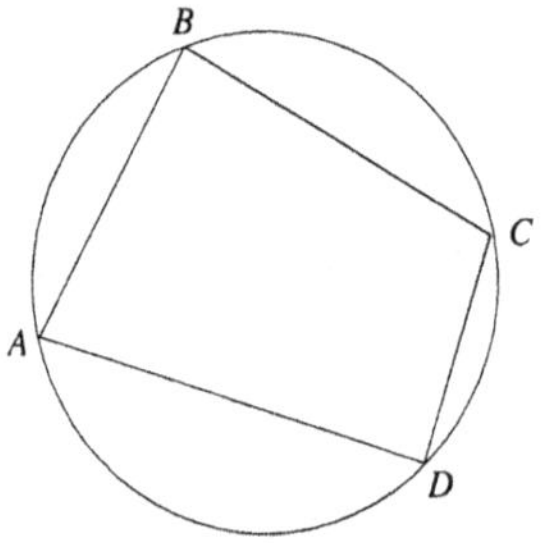

Note that we only have to show that $\angle A + \angle C = 180$, since if this is true, then

$$\angle B + \angle D = 360 - (\angle A + \angle C) = 360 - 180 = 180.$$

Suppose first that the quadrilateral $ABCD$ is cyclic. Draw the diagonals AC and BD and let P be the intersection of the diagonals, then use Thales' Theorem to get the angles as shown.

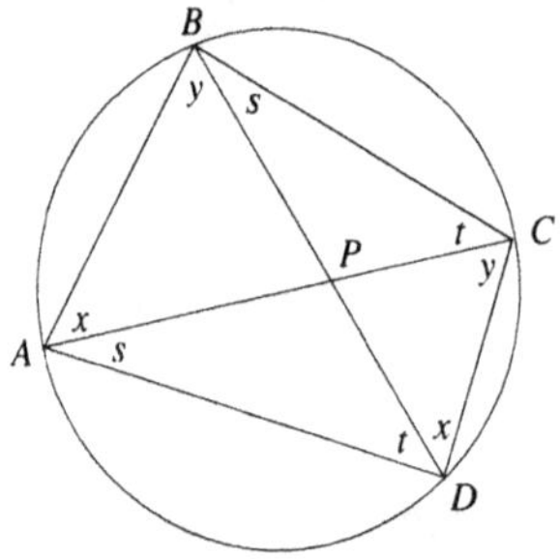

Since the sum of the internal angles in $\triangle ABC$ is 180, then

$$x + y + s + t = (x + s) + (y + t) = 180.$$

That is, $\angle A + \angle C = 180$ and $\angle B + \angle D = 180$, so that opposite angles are supplementary.

Conversely, suppose that $\angle A + \angle C = 180$ (and therefore that $\angle B + \angle D = 180$ also) and let the circle shown on the following page be the circumcircle of $\triangle ABC$.

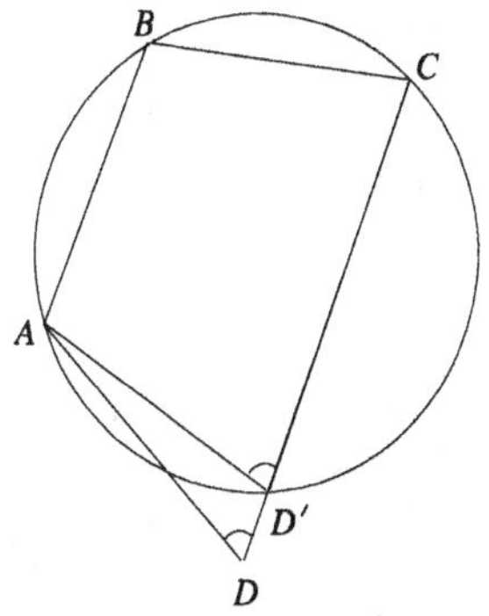

If the quadrilateral $ABCD$ is not cyclic, then the point D does not lie on this circumcircle. Assume that D lies outside the circle and let D' be the point where the line segment CD hits the circle. Since $ABCD'$ is a cyclic quadrilateral, $\angle B + \angle D' = 180$ and therefore $\angle D = \angle D'$, which contradicts the External Angle Inequality in $\triangle AD'D$.

If the point D is inside the circle, a similar argument leads to a contradiction of the External Angle Inequality.

Thus, if $\angle A + \angle C = 180$ and $\angle B + \angle D = 180$, then quadrilateral $ABCD$ is cyclic.

15. Given a circle $\mathcal{C}(P, s)$, a line l disjoint from $\mathcal{C}(P, s)$, and a radius r, $(r > s)$, construct a circle of radius r tangent to both $\mathcal{C}(P, s)$ and l.

    Note: The analysis figure indicates that there are four solutions.

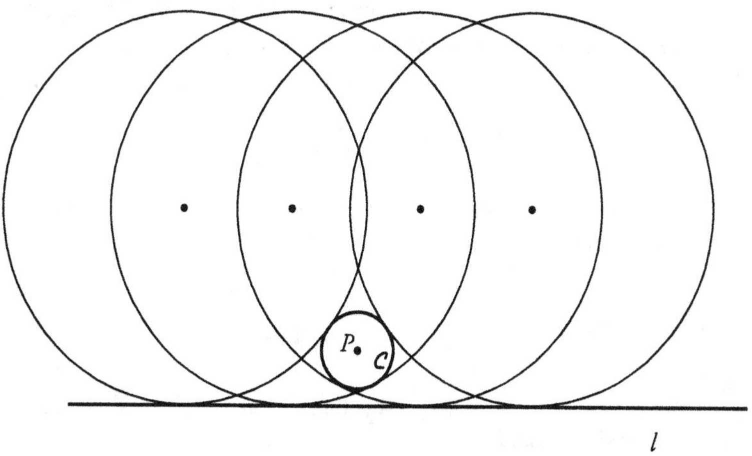

Solution. We see that the centers of the circles lie on the following constructible loci:

- a line parallel to l at distance r from l

- a circle $\mathcal{C}(P, r+s)$ or a circle $\mathcal{C}(P, r-s)$

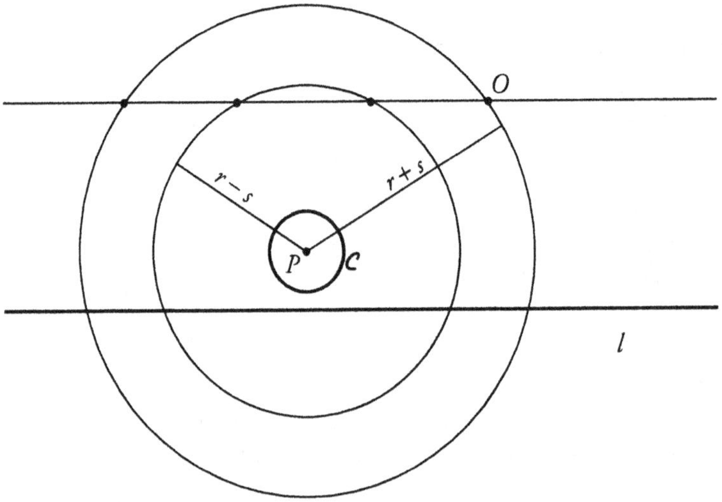

Since we are given the radius, the construction is reduced to finding the centers of the desired circles. We show how to construct one of them.

(1) Construct a line m parallel to l at distance r from l on the same side of l as $\mathcal{C}(P, s)$.

(2) Construct $\mathcal{C}(P, r+s)$.

(3) Let $O = m \cap \mathcal{C}(P, r+s)$. Note that if m and $\mathcal{C}(P, r+s)$ do not intersect, there is no solution.

(4) Construct $\mathcal{C}(O, r)$.

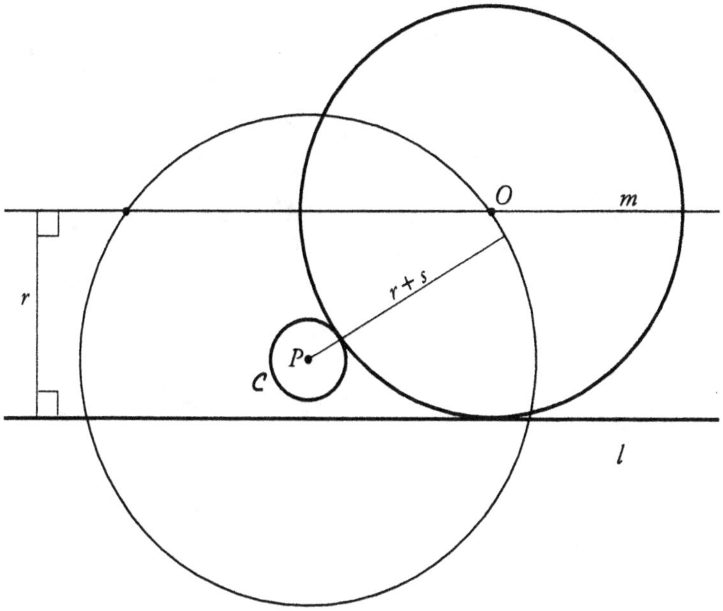
O
m
r
r + s
P
C
l

CHAPTER 2

CONCURRENCY

1. BE and CF are altitudes of $\triangle ABC$, and M is the midpoint of BC. Show that $ME \equiv MF$.

Solution. From the converse to Thales' Theorem, the circle with center M and diameter BC passes through the points E and F, since both $\angle BFC$ and $\angle BEC$ are right angles. Therefore, ME and MF are both radii of the circle and so $ME \equiv MF$.

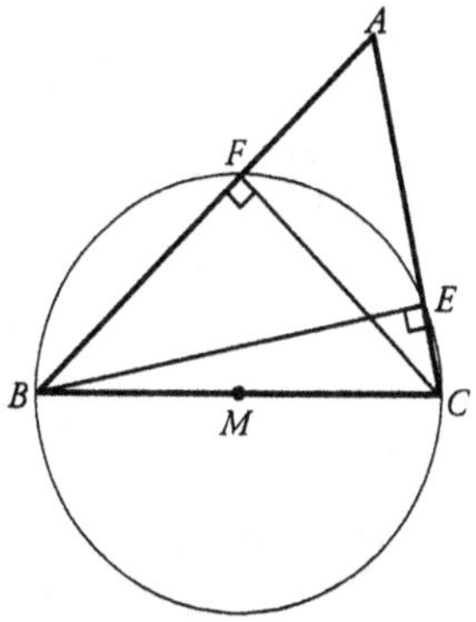

Solutions Manual to Accompany Classical Geometry: Euclidean, Transformational, Inversive, and Projective, First Edition. By I. E. Leonard, J. E. Lewis, A. C. F. Liu, G. W. Tokarsky.

3. The perpendicular bisector of side BC of $\triangle ABC$ meets the circumcircle at D on the opposite side of BC from A. Prove that AD bisects $\angle BAC$.

Solution. Referring to the figure, let O be the center of the circumcircle. The right bisector of BC passes through O. Note that $\triangle BOM \equiv \triangle COM$ by **HSR**, so that $\angle BOM = \angle COM$; that is, $\angle BOD = \angle COD$. By Thales' Theorem,

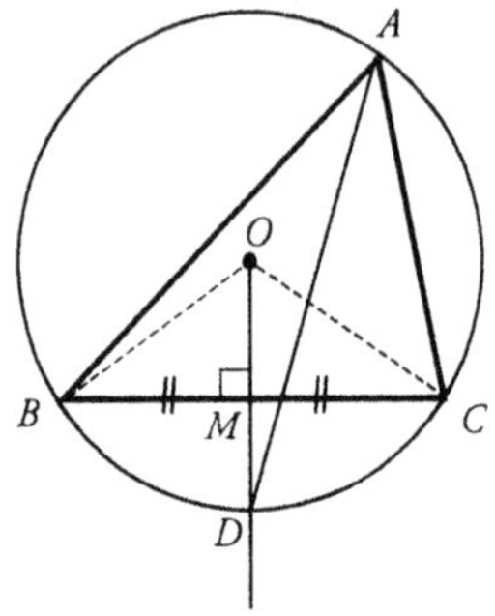

$$\begin{aligned}\angle BAD &= \tfrac{1}{2}\angle BOD \\ &= \tfrac{1}{2}\angle COD \\ &= \angle CAD,\end{aligned}$$

which shows that AD is the bisector of $\angle BAC$.

5. In the given figure, calculate the sizes of the angles marked α and β.

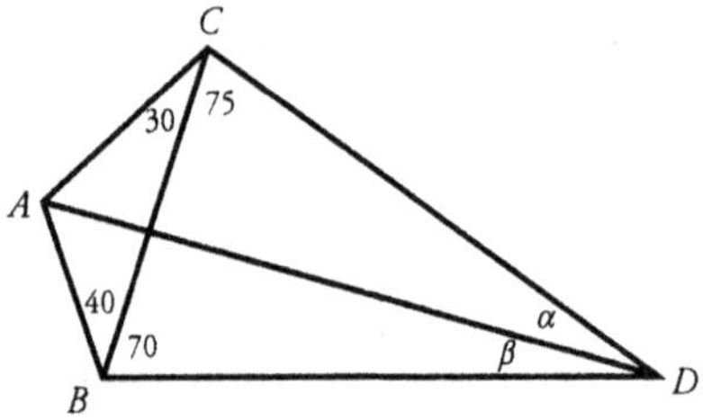

Solution. Note that for triangle ABC, the lines BD and CD are bisectors of the external angles $\angle B$ and $\angle C$, respectively.

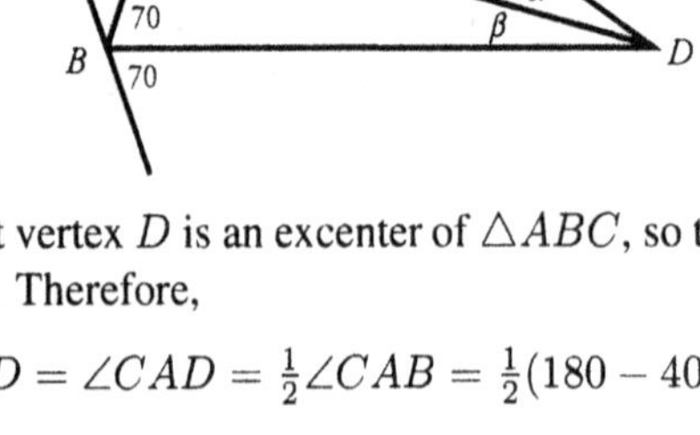

This means that vertex D is an excenter of $\triangle ABC$, so that AD is the internal bisector of $\angle A$. Therefore,

$$\angle BAD = \angle CAD = \tfrac{1}{2}\angle CAB = \tfrac{1}{2}(180 - 40 - 30) = 55,$$

so that

$$\alpha = 180 - 75 - 30 - 55 = 20 \quad \text{and} \quad \beta = 180 - 70 - 40 - 55 = 15.$$

7. Segments PS and PT are tangent to the circle at S and T. Show that

   (a) $PS \equiv PT$, and

   (b) $ST \perp OP$.

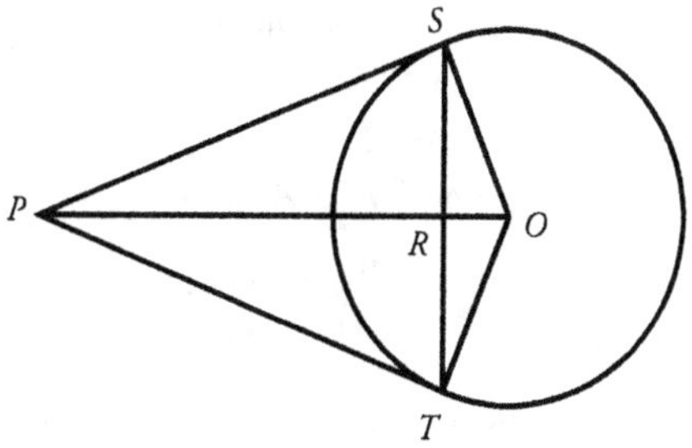

Solution.

(a) Since the radii SO and TO are perpendicular to the respective tangents, triangles PSO and PTO are right triangles with a common hypotenuse and with $SO = TO$. Thus, $\triangle PSO \equiv \triangle PTO$ by **HSR** and, consequently, $PS \equiv PT$.

(b) In triangles RSO and RTO we have

$$SO = TO,$$
$$\angle ROS \equiv \angle ROT \text{ (because } \triangle PSO \equiv \triangle PTO\text{)},$$
$$RO \text{ is common,}$$

so $\triangle RSO \equiv \triangle RTO$ by **SAS**. Thus, $\angle SRO = \angle TRO$ and, since the sum of the angles is 180, it follows that $ST \perp OP$.

9. Construct $\triangle ABC$ given the side BC, the length h_b of the altitude from B, and the length m_a of the median from A.

Solution.

Analysis Figure. The expected solution is shown below.

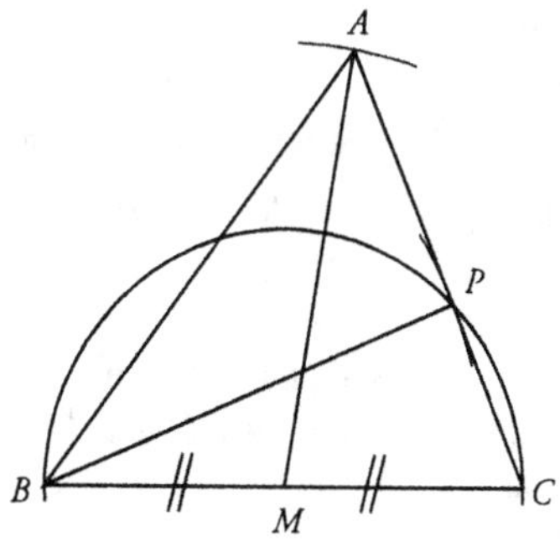

Construction.

(1) On BC, construct a semicircle with BC as diameter and the midpoint M of BC as center. The construction lines have been omitted in the diagram.

(2) With center B and radius equal to h_b, draw an arc cutting the semicircle at P so that PB is the altitude.

(3) Draw the ray $\overrightarrow{CP}$. With center M and radius equal to h_c, draw an arc cutting CP at A. $\triangle ABC$ is the desired triangle.

Justification. $\angle BPC$ is a right angle by Thales' Theorem, so BP is an altitude of length h_b. AM is a median and is of length m_a.

11. Construct triangle ABC given BC, an angle β congruent to $\angle B$, and the length t of the median from B.

Solution.

Analysis Figure. The expected solution is shown below.

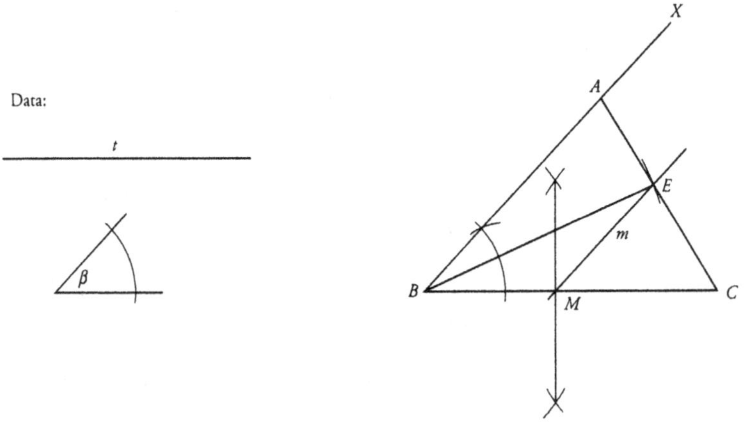

Construction.

(1) Copy angle β, forming $\angle XBC$.

(2) Bisect BC, obtaining the midpoint M.

(3) Construct a line m through M parallel to BX.

(4) With center B and radius t, draw an arc cutting m at E.

(5) Draw the line through C and E cutting BX at A. $\triangle ABC$ is the desired triangle.

Justification. We need to show that E is the midpoint of AC. This follows from the fact that M is the midpoint of BC and $ME \parallel AB$ (see the converse of the Midline Theorem).

13. Given segments AB and CD, which meet at a point P off the page, construct the bisector of $\angle P$. All constructions must take place within the page.

Solution.

Analysis Figure. The expected solution is shown below.

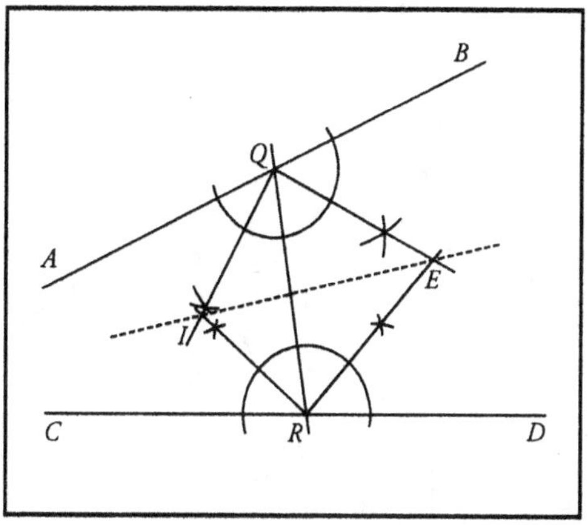

Construction.

(1) Draw a transversal PQ for AB and CD.

(2) Bisect angles BQR and DRQ and let the bisectors meet at point E. Referring to the diagram, E would be an excenter of $\triangle PQR$, so E is on the interior bisector of $\angle QPR$.

(3) Bisect angles AQR and CRQ and let the bisectors meet at point I. Then I would be the incenter of $\triangle PQR$, so I is on the interior bisector of $\angle QPR$.

(4) Draw the line IE. This is the bisector of $\angle QPR$.

Justification. This is left as an exercise for the reader.

15. M is the midpoint of the chord AB of a circle $\mathcal{C}(O, r)$. Show that if a different chord CD contains M, then $AB < CD$. (You may use Pythagoras' Theorem.)

Solution. Let O be the center of the circle, let r be its radius, and let N be the midpoint of CD, then $OM \perp AB$ and $ON \perp CD$. OM is the hypotenuse of ΔOMN, so that $ON < OM$. Using Pythagoras' Theorem for both triangles OCN and OBM, we have

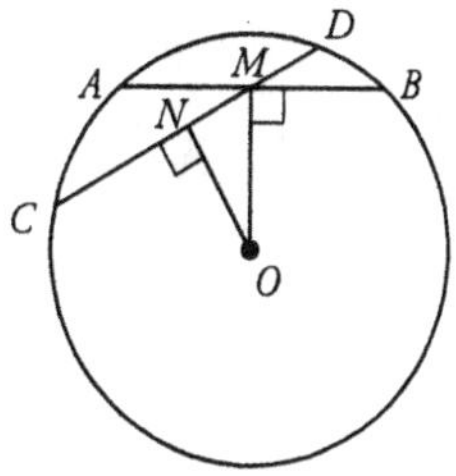

$$ON^2 + NC^2 = r^2 = OM^2 + MB^2,$$

and since $ON^2 < OM^2$, it follows that

$$NC^2 > MB^2,$$

and hence

$$CD = 2NC > 2MB = AB.$$

CHAPTER 3

SIMILARITY

1. P and Q are points on the side BC of $\triangle ABC$ with $BP = PQ = QC$. The line through P parallel to AC meets AB at X, and the line through Q parallel to AB meets AC at Y. Show that $\triangle ABC \sim \triangle AXY$.

Solution. In the figure below, let Z be the point where PX and QY meet.

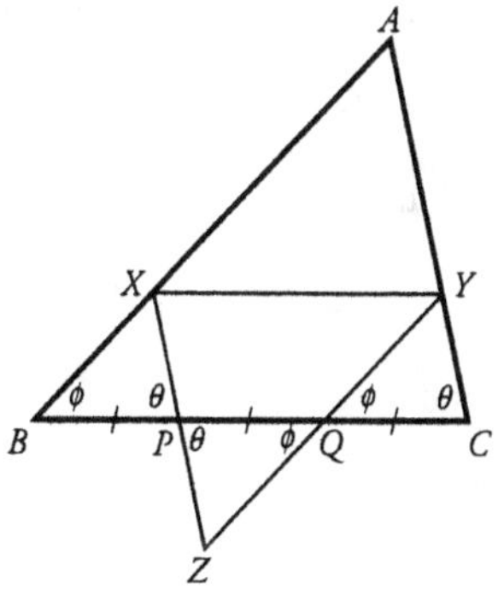

Solutions Manual to Accompany Classical Geometry: Euclidean, Transformational, Inversive, and Projective, First Edition. By I. E. Leonard, J. E. Lewis, A. C. F. Liu, G. W. Tokarsky.

Note that $\angle YCQ \equiv \angle XPB$ because $PX \parallel AC$ and that $\angle XPB \equiv QPZ$ because they are vertically opposite angles. Thus, $\angle YCQ \equiv \angle QPZ$ so that in triangles QPZ and QCY we have

$$\begin{aligned} \angle QCY &\equiv \angle QPZ, \\ QC &\equiv QP, \\ \angle CQY &\equiv \angle PQZ \quad \text{(vertically opposite angles).} \end{aligned}$$

Therefore, $\triangle QCY \equiv \triangle QPZ$ by **ASA** and so $QZ \equiv QY$; that is, Q is the midpoint of YZ.

Similarly, P is the midpoint of XZ. It follows that $PQ \parallel XY$; that is, $BC \parallel XY$. From this it follows that $\triangle AXY \sim \triangle ABC$.

Alternate Solution. Since XP is parallel to AC, then $\triangle XBP \sim \triangle ABC$, and since YQ is parallel to AB, then $\triangle YQC \sim \triangle ABC$.

Therefore,

$$\frac{AB}{XB} = \frac{BC}{BP} = 3,$$

and

$$AX = AB - XB = 3XB - XB = 2XB,$$

so that

$$\frac{AX}{AB} = \frac{2XB}{3XB} = \frac{2}{3}.$$

Similarly,

$$\frac{AC}{YC} = \frac{BC}{QC} = 3,$$

and

$$AY = AC - YC = 3YC - YC = 2YC,$$

so that

$$\frac{AY}{AC} = \frac{2YC}{3YC} = \frac{2}{3}.$$

Since $\angle XAY \equiv \angle BAC$, then

$$\triangle AXY \sim \triangle ABC$$

by the **sAs** similarity condition.

3. In the figure on the right, $\triangle ABC$ is an isosceles triangle with altitude AD,

$$AB = \tfrac{3}{2}BC, \quad AF = 4FD, \quad \text{and} \quad FD = 1.$$

FE is a perpendicular from F to AC. Find the length of FE.

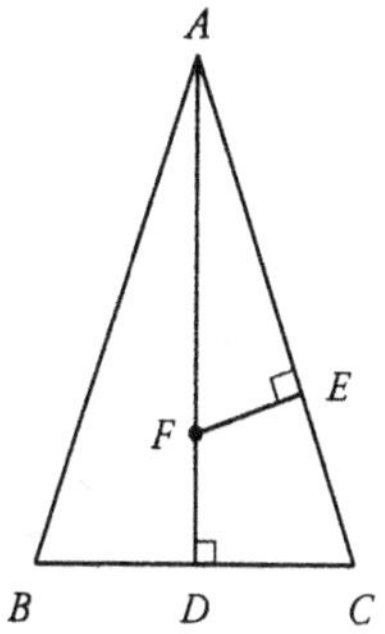

Solution. From the figure, since AD is an altitude, AD is perpendicular to BC and

$$\angle ADB = \angle ADC = 90.$$

Now, side AD is common to both $\triangle ADB$ and $\triangle ADC$, while AB and AC are hypotenuses of the right triangles $\triangle ADB$ and $\triangle ADC$, respectively. Therefore, by the **HSR** congruency theorem, $\triangle ADB \equiv \triangle ADC$ so that $BD = DC$ and $DC = \frac{1}{2}BC$.

Since $\angle CAD$ is common to the two right triangles AEF and ADC, then $\triangle AEF \sim \triangle ADC$ by the **AA** similarity theorem, so that

$$\frac{AF}{AC} = \frac{FE}{DC} = \frac{FE}{\frac{1}{2}BC} = \frac{2FE}{BC},$$

and since $AF = 4FD = 4 \cdot 1 = 4$, then

$$\frac{4}{AC} = 2\frac{FE}{BC},$$

and

$$FE = 2\frac{BC}{AC} = 2 \cdot \frac{\frac{2}{3}AB}{AB} = \frac{4}{3}.$$

5. The bisector of $\angle BAC$ meets BC at D. The circle with center C passing through D meets AD at X. Prove that $AB \cdot AX = AC \cdot AD$.

Solution. Let $\alpha = \angle CXD$, $\beta = \angle CDX$, and $\gamma = \angle XCD$. Note that in the figure on the following page, $\alpha = \beta$ since $\triangle CXD$ is isosceles.

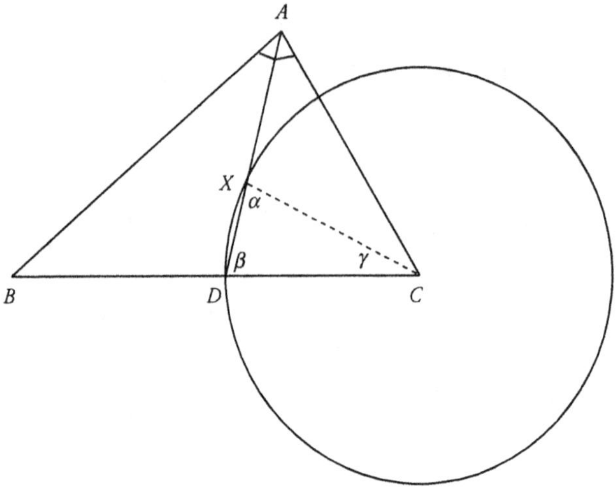

From the Exterior Angle Theorem, we have

$$\angle BDA = \alpha + \gamma = \beta + \gamma$$

and

$$\angle AXC = \beta + \gamma.$$

In $\triangle ABD$ and $\triangle ACX$, we have $\angle BAD = \angle CAX$ and $\angle BDA = \angle CXA$. By the **AA** similarity condition, this means that $\triangle ABD \sim \triangle ACX$. Therefore,

$$\frac{AB}{AC} = \frac{AD}{AX}.$$

That is, $AB \cdot AX = AC \cdot AD$.

7. Given a triangle ABC with acute angles B and C, construct a square $PQRS$ with PQ in BC and vertices R and S in AB and AC, respectively.

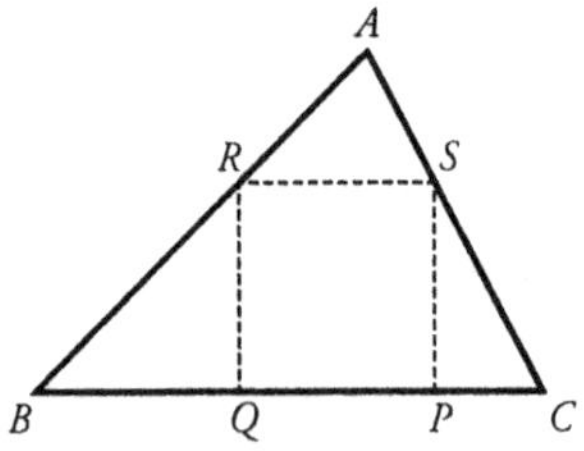

Solution.

Analysis Figure. Note that construction arcs for standard constructions have been omitted in the diagram.

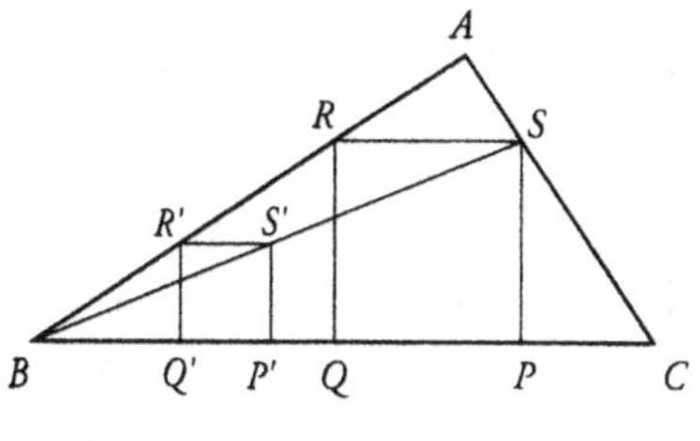

Construction.

(1) From a point R' on AB, drop a perpendicular $R'Q'$ to BC.

(2) Complete the square $P'Q'R'S'$ with edge length $R'Q'$ and with $P'Q'$ in BC.

(3) Draw the line BS' meeting AC at S. Drop the perpendicular SP to BC.

(4) Draw the line RS parallel to BC meeting AB at R. Note that $RS \perp SP$.

Justification. Since $QP \parallel RS$, we have $\angle SPQ = 180 - \angle PSQ = 90$, and so PQ is in side BC of the triangle. It follows also that $PQRS$ is a rectangle, and it remains to show that $PQRS$ is a square.

Since both $R'S'$ and RS are parallel to BC, they are parallel to each other. Thus, triangles BRS and $BR'S'$ are similar, so that

$$\frac{RS}{R'S'} = \frac{BS}{BS'}.$$

Also, $S'P'$ and SP are parallel, so triangles BPS and $BP'S'$ are similar and thus

$$\frac{PS}{P'S'} = \frac{BS}{BS'}.$$

It follows that

$$\frac{PS}{P'S'} = \frac{RS}{R'S'}$$

and, since $P'S' = R'S'$, we get $PS = RS$; that is, the rectangle $PQRS$ is a square.

Alternate Solution.

Analysis Figure. The expected solution is shown on the following page.

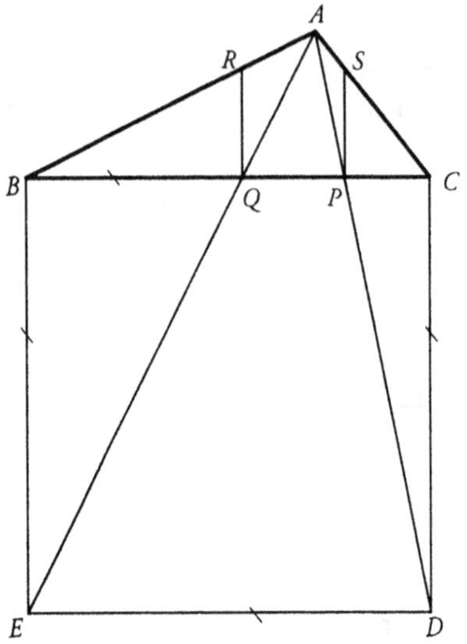

Construction.

(1) Construct the square $BCDE$ on side BC, as shown.

(2) Draw lines AD and AE cutting BC at P and Q, respectively.

(3) Draw PS parallel to DC cutting AC at S.

(4) Draw QR parallel to EB cutting AB at R.

Then $PQRS$ is the desired square.

Justiication. First note that $QR \parallel EB$, so $QR \perp BC$. Similarly, $PS \perp BC$. Thus, it suffices to show that $QR = QP = PS$, since this will imply that $PQRS$ is a square.

Since $QR \parallel EB$, triangles AQR and AEB are similar, with

$$\frac{QR}{EB} = \frac{AQ}{AE}.$$

Since $PS \parallel DC$, triangles APS and ADC are similar, with

$$\frac{PS}{DC} = \frac{AP}{AD}.$$

Since $QP \parallel ED$, triangles AQP and AED are similar, with

$$\frac{AQ}{AE} = \frac{QP}{ED} \quad \text{and} \quad \frac{AP}{AD} = \frac{QP}{ED}.$$

Thus,

$$\frac{QR}{EB} = \frac{QP}{ED} = \frac{PS}{DC},$$

and since $EB = ED = DC$, we are finished.

9. Given two disjoint circles $\mathcal{C}(O, R)$ and $\mathcal{C}(Q, r)$, with $R > r$, construct the two "internal" tangents:

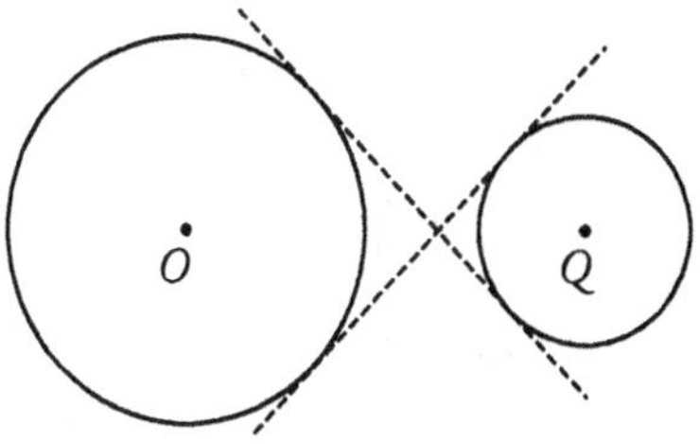

Solution.

Analysis Figure. The expected solution is shown below.

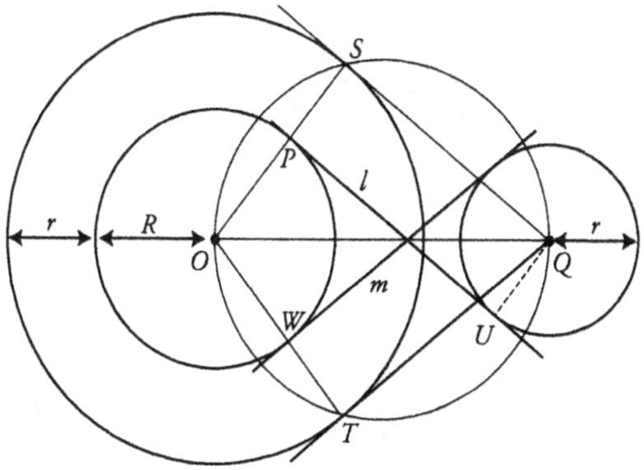

Construction.

(1) Draw $\mathcal{C}(O, R + r)$.

(2) Draw the tangents QS and QT to $\mathcal{C}(O, R + r)$.

(3) Draw the segments OS and OT intersecting $\mathcal{C}(O, R)$ at P and U, respectively.

(4) Draw a line l through P parallel to QS. Draw a line m through R parallel to QT. These are the desired tangent lines.

Justification. Let QU be parallel to SP so that $QSPU$ is a rectangle. Then $QU = PS = r$, where r is the radius of the smaller circle and, since $PU \perp QU$, PU is tangent to the smaller circle. Since OP is a radius of the larger circle and, since $PU \perp OP$, then PU is tangent to the larger circle; that is, l is tangent to both circles. Similar reasoning shows that m is tangent to both circles.

11. In the diagram, $PA = 3$, $BC = 4$ and $PC = 6$. Find the length of the segment AB.

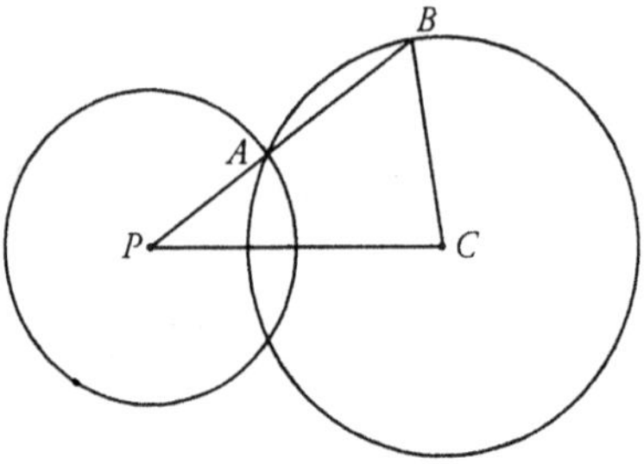

Solution. Let r be the radius of the larger circle. By the power of the point P with respect to the larger circle,

$$PA \cdot PB = PC^2 - r^2,$$

so that

$$3 \cdot PB = 6^2 - 4^2,$$

and thus

$$PB = \frac{20}{3}.$$

Hence,

$$AB = PB - PA = \frac{11}{3}.$$

13. In the following diagram, the segment AB is of length 3. Construct all points on the line AB whose power with respect to ω is 4.

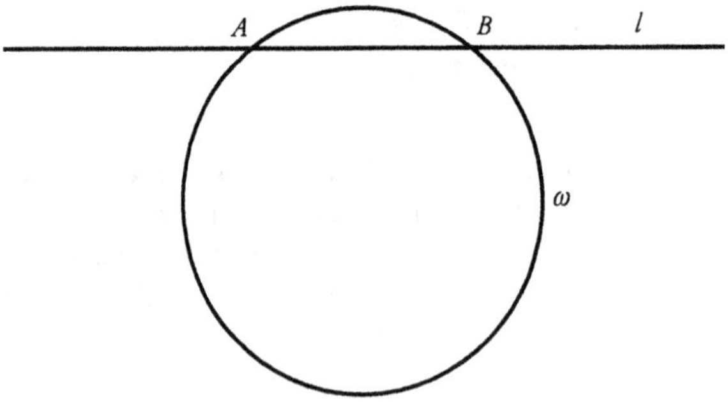

Solution. If a point P is on the line outside the circle then $\overline{PA} \cdot \overline{PB} = 4$, so

$$\overline{PA} \cdot (\overline{PA} + 3) = 4.$$

Solving this quadratic for $\overline{PA}$, we get

$$\overline{PA} = 1 \quad \text{or} - 4.$$

Hence, there are two points on the line AB and they can be constructed as follows:

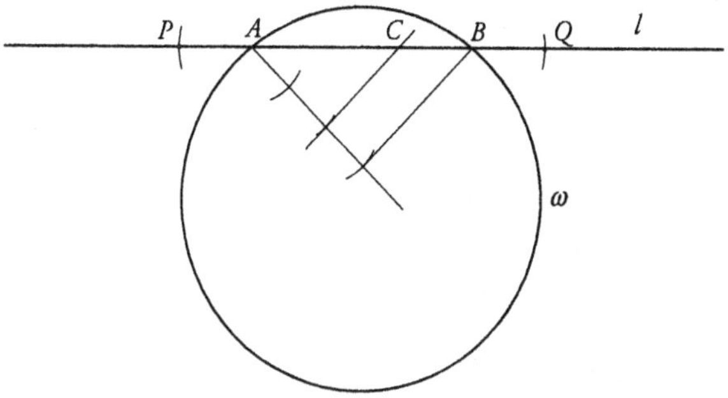

(1) Trisect the segment AB in the usual way to obtain a segment BC of length 1.

(2) With center A and radius BC, draw an arc cutting the line AB at a point P outside the segment AB.

(3) With center B and radius BC, draw an arc cutting the line AB at a point Q outside the segment AB. P and Q are the desired points.

15. Two different circles intersect at two points A and B. Find all points P such that the power of P is the same with respect to both circles.

Solution. The circles are shown in the figure below.

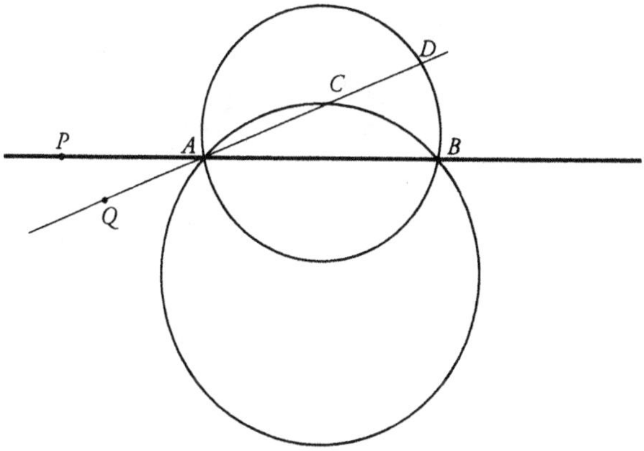

For each point P on the line AB, the power of P with respect to both circles is $\overline{PA} \cdot \overline{PB}$.

If Q is a point not on the line AB, draw the line QA. It intersects one circle at C and the other at D with $D \neq C$. Thus, $\overline{QC} \neq \overline{QD}$ and $\overline{QA} \cdot \overline{QC} \neq \overline{QA} \cdot \overline{QD}$. Therefore, the power of Q with respect to the circles is different. This shows that the set of points P such that the power of the point P is the same with respect to both circles is the line AB.

CHAPTER 4

THEOREMS OF CEVA AND MENELAUS

1. Show that the lines drawn from a vertex to a point halfway around the perimeter of a triangle are concurrent.

   This point of concurrency is called the ***Nagel point*** of the triangle. In the diagram, a, b, and c denote the lengths of the sides.

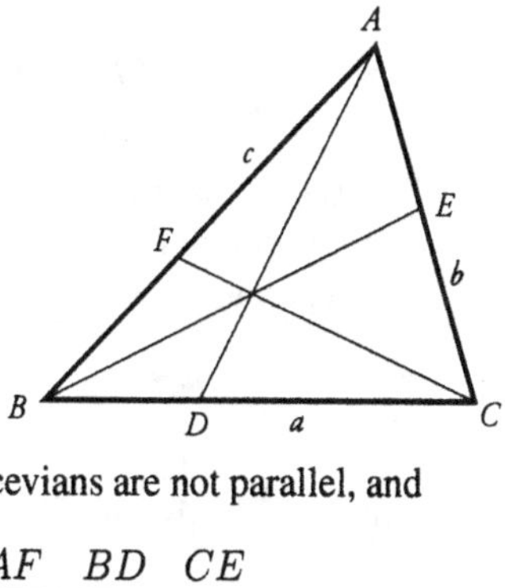

Solution. Since all cevians are internal, the cevians are not parallel, and

$$\frac{\overline{AF}}{\overline{FB}} \cdot \frac{\overline{BD}}{\overline{DC}} \cdot \frac{\overline{CE}}{\overline{EA}} = \frac{AF}{FB} \cdot \frac{BD}{DC} \cdot \frac{CE}{EA}.$$

Letting $s = \frac{1}{2}(a + b + c)$, then

$$\begin{aligned} AF &= s - b, & BD &= s - c, & CE &= s - a, \\ FB &= s - a, & DC &= s - b, & EA &= s - c. \end{aligned}$$

Solutions Manual to Accompany Classical Geometry: Euclidean, Transformational, Inversive, and Projective, First Edition. By I. E. Leonard, J. E. Lewis, A. C. F. Liu, G. W. Tokarsky.

Therefore,

$$\frac{\overline{AF}}{\overline{FB}} \cdot \frac{\overline{BD}}{\overline{DC}} \cdot \frac{\overline{CE}}{\overline{EA}} = \frac{s-b}{s-a} \cdot \frac{s-c}{s-b} \cdot \frac{s-a}{s-c} = 1\,,$$

and by Ceva's Theorem the lines are concurrent.

3. Prove the necessary part of Ceva's Theorem for parallel cevians; that is, prove the following:

If the cevians AD, BE, and CF are parallel cevians for triangle ABC, then

$$\frac{\overline{AF}}{\overline{FB}} \cdot \frac{\overline{BD}}{\overline{DC}} \cdot \frac{\overline{CE}}{\overline{EA}} = 1.$$

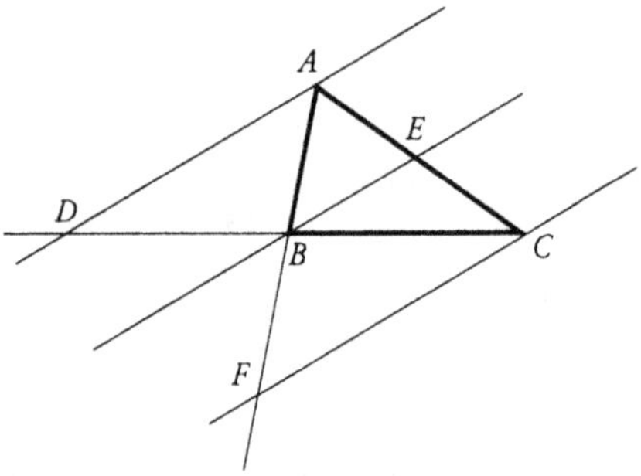

Solution. Referring to the diagram, because the lines AD and EB are parallel, $\triangle ADC \sim \triangle EBC$, so that

$$\frac{AE}{EC} = \frac{DB}{BC}.$$

Since the lines AD and CF are parallel, $\triangle ABD \sim \triangle FBC$, so

$$\frac{AF}{FB} = \frac{CD}{BC}.$$

Consequently,

$$\frac{AF}{FB} \cdot \frac{BD}{DC} \cdot \frac{CE}{EA} = \frac{CD}{BC} \cdot \frac{BD}{DC} \cdot \frac{BC}{BD} = 1.$$

This shows that the magnitude of the cevian product is 1. Referring to the diagram, two of the cevians are external and one is internal, so the sign of the cevian product is positive, and

$$\frac{\overline{AF}}{\overline{FB}} \cdot \frac{\overline{BD}}{\overline{DC}} \cdot \frac{\overline{CE}}{\overline{EA}} = +1.$$

5. Prove that AD is an angle bisector in the figure below.

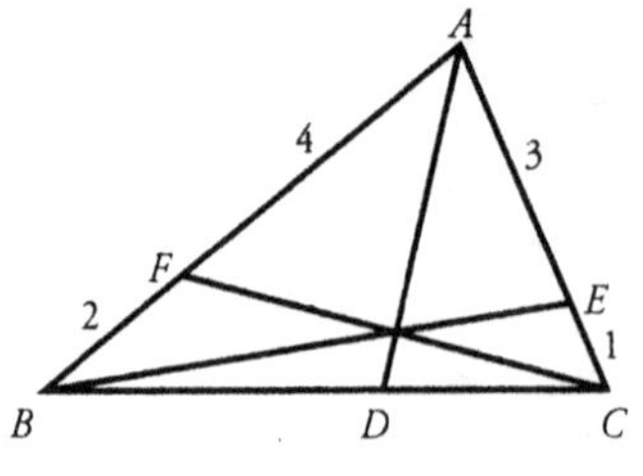

Solution. From Ceva's Theorem, we have

$$\frac{\overline{AF}}{\overline{FB}} \cdot \frac{\overline{BD}}{\overline{DC}} \cdot \frac{\overline{CE}}{\overline{EA}} = 1,$$

which implies that

$$\frac{4}{2} \cdot \frac{BD}{DC} \cdot \frac{1}{3} = 1,$$

and thus we get

$$\frac{BD}{DC} = \frac{3}{2},$$

that is,

$$\frac{DB}{DC} = \frac{AB}{AC}.$$

It now follows from the converse to the Angle Bisector Theorem that AD is the angle bisector.

7. (a) Prove that an external angle bisector is parallel to the opposite side if and only if the triangle is isosceles.
   (b) Show that in a nonisosceles triangle, the external angle bisectors meet the opposite sides in three collinear points.

Solution.

(a) Suppose that the triangle is isosceles. Let m be the exterior angle bisector of the angle opposite the base. Referring to the figure below, we have $\gamma = \delta$ since the triangle is isosceles. We have $\alpha = \beta$ because m is an angle bisector. Thus, by the Exterior Angle Theorem,

$$2\alpha = \alpha + \beta = \gamma + \delta = 2\gamma.$$

It follows that $\alpha = \gamma$, and hence m is parallel to the opposite side.

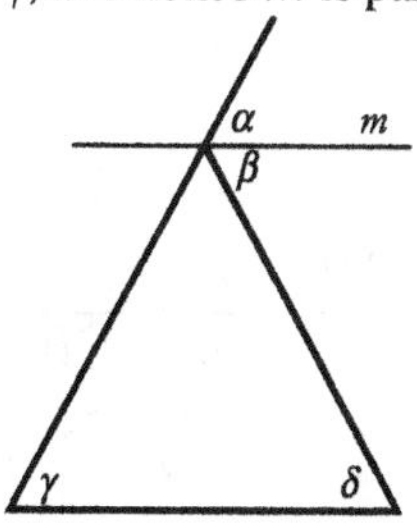

Conversely, suppose m is an external angle bisector that is parallel to the opposite side. Referring to the figure, this means that $\gamma = \alpha = \beta = \delta$, so the triangle has two congruent angles and is therefore isosceles.

(b) Since the triangle is not isosceles, each exterior angle bisector must intersect the opposite side by part (a) above. Assuming that the triangle is ABC and the bisectors are AD, BE, and CF, as in the figure below, we need to show that

$$\frac{\overline{AF}}{\overline{FB}} \cdot \frac{\overline{BD}}{\overline{DC}} \cdot \frac{\overline{CE}}{\overline{EA}} = -1.$$

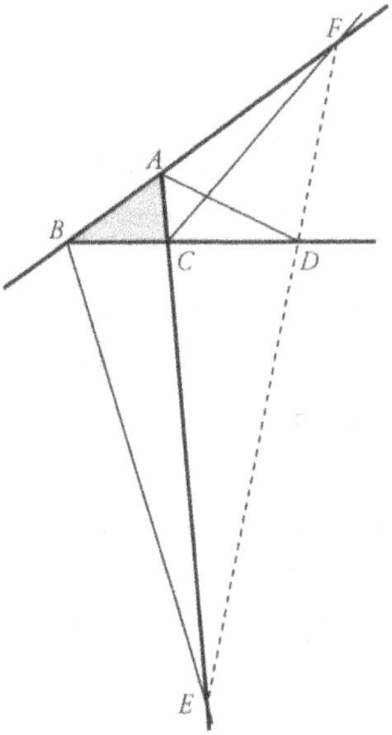

Since all exterior angle bisectors divide the opposite sides externally, the sign of each of the directed ratios is negative, so the cevian product is negative and we need only show that the magnitude is 1, that is, that

$$\frac{AF}{FB} \cdot \frac{BD}{DC} \cdot \frac{CE}{EA} = 1.$$

From the Angle Bisector Theorem, for external angles we have

$$\frac{AF}{FB} = \frac{AC}{CB}, \qquad \frac{BD}{DC} = \frac{BA}{AC}, \qquad \frac{CE}{EA} = \frac{CB}{BA},$$

so that

$$\frac{AF}{FB} \cdot \frac{BD}{DC} \cdot \frac{CE}{EA} = \frac{AC}{CB} \cdot \frac{BA}{AC} \cdot \frac{CB}{BA} = 1.$$

Therefore, by Menelaus' Theorem, the points D, E, and F are collinear.

CHAPTER 5

AREA

1. In the regular octagon shown below, is the area of the shaded region larger or smaller than the total area of the unshaded regions?

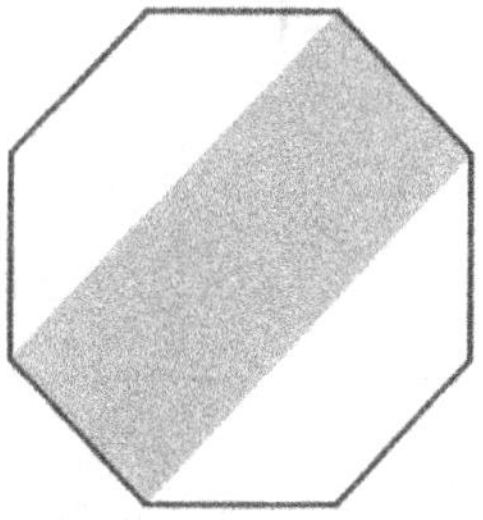

Solution. In the figure on the following page, the shaded and unshaded regions have been partitioned into smaller congruent regions, and we can see that the areas of the shaded and unshaded regions are the same size.

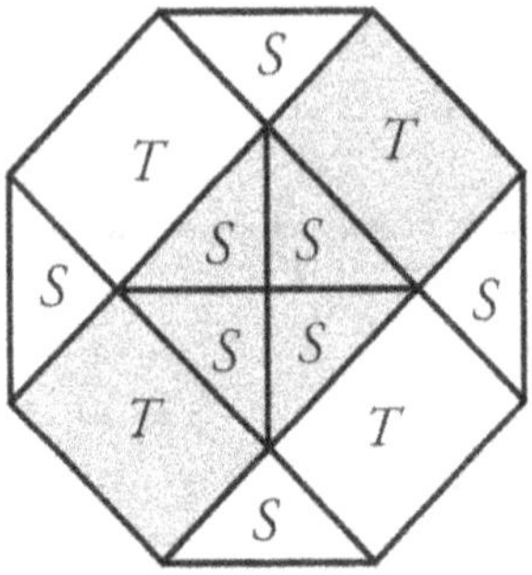

3. A square is divided into five nonoverlapping rectangles, with four of the rectangles completely surrounding the fifth rectangle, as shown in the diagram. The outer rectangles are the same area. Prove that the inner rectangle is a square.

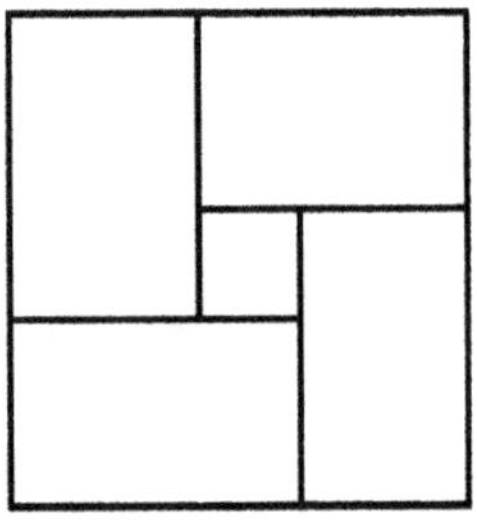

Solution. After labelling the edges of the rectangles, as shown below, if we can show that $a = c = e = g$ (and hence that $b = d = f = h$), then it will follow that $x = g - d$ and $y = a - f = g - d$, thus showing that the inner rectangle is a square.

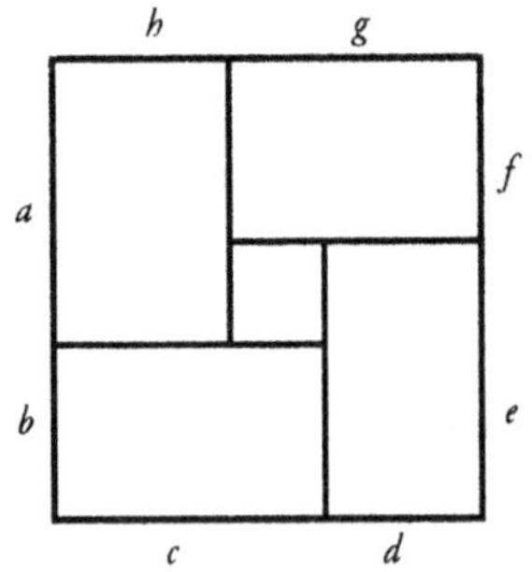

Suppose that $a \neq c$, then either $a > c$ or $a < c$. We may assume that $a > c$ (the proof for the case where $a < c$ can be obtained by relabelling the diagram).

Since $a + b = c + d$ and $a > c$, we get $b < d$.

Since $bc = de$ and $b < d$, we get $c > e$.

Since $c + d = e + f$ and $c > e$, we get $d < f$.

Continuing in this fashion, we get $e > g$, $f < h$, and $g > a$. However, this implies that $a > c > e > g > a$, which is a contradiction. Therefore, our assumption that $a \neq c$ is incorrect; that is, we must have $a = c$.

Similarly, $c = e$, $e = g$, and $g = a$, finishing the proof.

5. A paper rectangle $ABCD$ of area 1 is folded along a straight line so that C coincides with A. Prove that the area of the pentagon obtained is less than $\frac{3}{4}$.

Solution. In the figure below, the rectangle is folded along the line PQ, which is the right bisector of the diagonal AC.

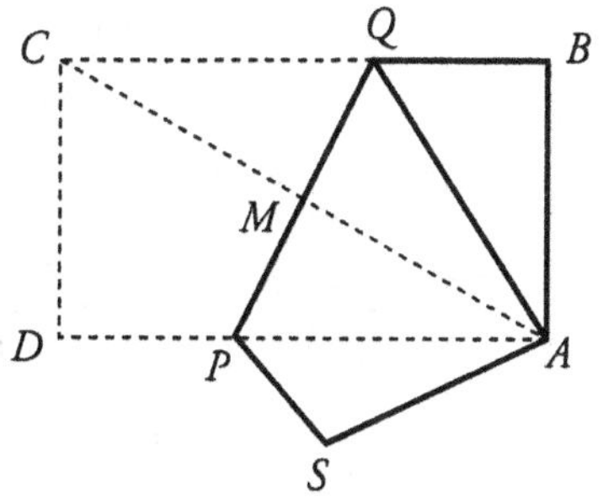

Since the hypotenuse is the longest side of a triangle, we have

$$AP > AM = \tfrac{1}{2}AC > \tfrac{1}{2}AD,$$

so that

$$[APQ] = \tfrac{1}{2}AP \cdot AB > \tfrac{1}{2}\left(\tfrac{1}{2}AD\right)AB;$$

that is,

$$[APQ] > \tfrac{1}{4}.$$

Now,

$$\begin{aligned}[ABQPS] &= [AQPS] + [ABQ] \\ &= [CQPD] + [ABQ] \\ &= [ABCD] - [APQ] \\ &< 1 - \tfrac{1}{4} \\ &= \tfrac{3}{4}.\end{aligned}$$

7. $ABCDEF$ is a convex hexagon in which opposite sides are parallel. Prove that $[ACE] = [BDF]$.

Solution. In the figure below, let

h_1 be the perpendicular distance between the parallel sides BC and EF,
h_2 be the perpendicular distance between the parallel sides AB and ED, and
h_3 be the perpendicular distance between the parallel sides AF and CD.

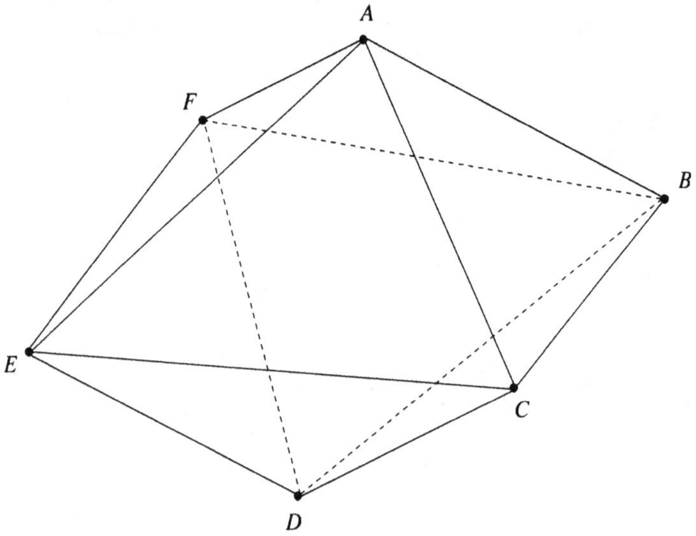

We have

$$\begin{aligned}[AFE] + [ABC] &= h_1(EF + BC)/2 \\ [CBA] + [CED] &= h_2(AB + ED)/2 \\ [AEF] + [CDE] &= h_3(AF + CD)/2,\end{aligned}$$

that is,

$$\begin{aligned}4[AEF] + 4[ABC] + 4[CDE] = h_1(EF + BC) &+ \\ h_2(AB + ED) + h_3(AF + CD) & \quad (*)\end{aligned}$$

Similarly,

$$\begin{aligned}[DFE] + [DCB] &= h_1(EF + BC)/2 \\ [AFB] + [FED] &= h_2(AB + DE)/2 \\ [AFB] + [BDC] &= h_3(AF + CD)/2,\end{aligned}$$

that is,

$$\begin{aligned}4[AFB] + 4[FED] + 4[BCD] = h_1(EF + BC) &+ \\ h_2(AB + ED) + h_3(AF + CD) & \quad (**)\end{aligned}$$

From $(*)$ and $(**)$, we have

$$[ABC] + [AEF] + [CDE] = [AFB] + [FED] + [BCD]$$

so that

$$\begin{aligned}[ACE] &= [ABCDEF] - ([ABC] + [AEF] + [CDE]) \\ &= [ABCDEF] - ([AFB] + [FED] + [BCD]) \\ &= [BDF].\end{aligned}$$

9. Seven children share a square pizza whose crust may be considered to consist only of the perimeter. Show how they make straight cuts to divide the pizza into seven pieces such that all pieces have the same amount of pizza and the same amount of crust.

Solution. Let P be the center of the square (where the diagonals intersect), as in the figure on the following page, and let l be the total length of the perimeter.

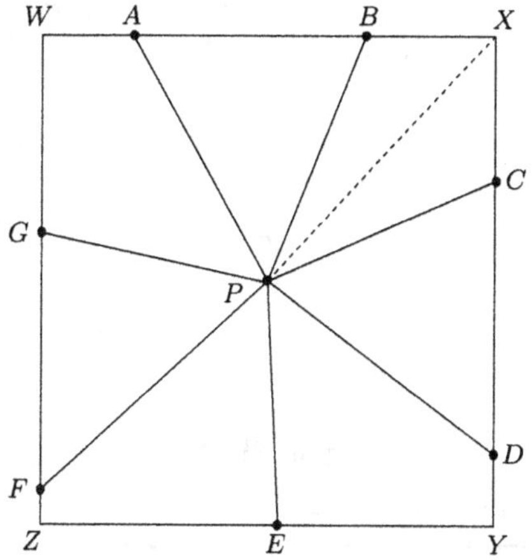

Starting anywhere on the perimeter, mark seven points A, B, C, D, E, F, and G that divide the perimeter into seven equal lengths. The cuts are AP, BP, CP, DP, EP, FP, and GP. The length of the crust of each piece is $l/7$. The area of each piece is the same since P is at distance $l/8$ from all four sides of the square. Thus, all triangular pieces, like ABP, have the same area, namely, $l/112$. Also, all quadrilateral pieces (like $BXCP$) are composed of two triangular pieces whose bases (BX and XC) have a total length of $l/7$ and whose altitudes are $l/8$.

11. $ABCDEF$ is a convex hexagon with side AB parallel to CF, side CD parallel to BE, and side EF parallel to AD. Prove that $[ACE] = [BDF]$.

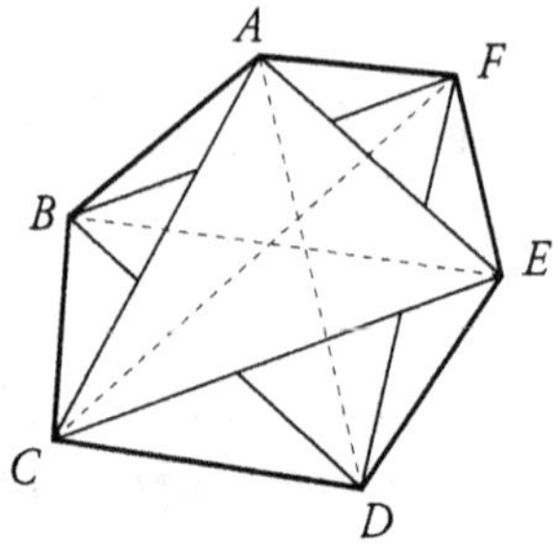

Solution. We have the following:

$$[ABC] = [ABF] \qquad \text{(because } AB \parallel CF)$$
$$[CDE] = [CDB] \qquad \text{(because } CD \parallel BE)$$
$$[EFA] = [EFD] \qquad \text{(because } EF \parallel AD).$$

Therefore,

$$\begin{aligned}[ACE] &= [ABCDEF] - ([ABC] + [CDE] + [EFA]) \\ &= [ABCDEF] - ([ABF] + [CDB] + [EFD]) \\ &= [BDF].\end{aligned}$$

13. Given a convex quadrilateral $ABCD$, construct a point E on the extension of BC such that the area $[ABCD] = [ABE]$.

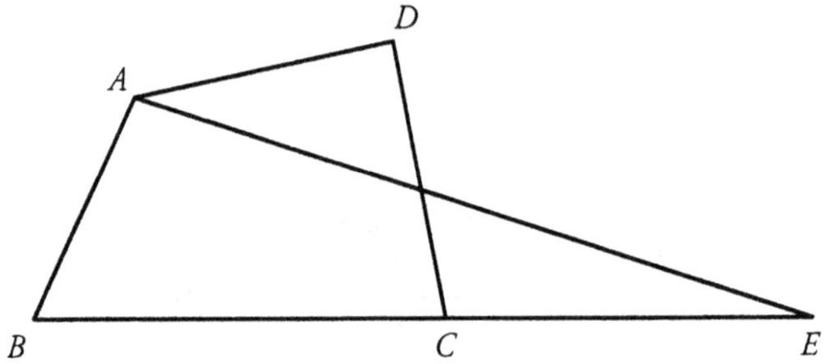

Solution. Draw DE parallel to AC, hitting BC at E, as in the figure below.

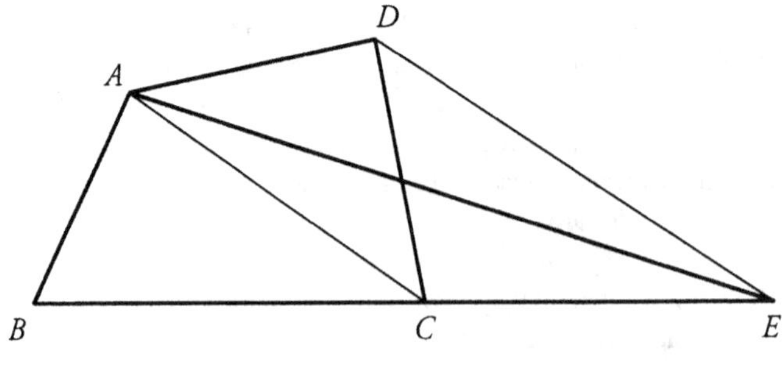

Then $\triangle ACD$ and $\triangle ACE$ both have base AC. Since AC and DE are parallel, they also have the same height, so that $[ACE] = [ACD]$.

Since $\triangle ACD$ and $\triangle ACE$ both contain $\triangle ACK$, then

$$\begin{aligned}[ADK] &= [ACD] - [ACK] \\ &= [ACE] - [ACK] \\ &= [CEK].\end{aligned}$$

Therefore, by the additivity of the area function,

$$\begin{aligned}[ABCD] &= [ABC] + [ADC] \\ &= [ABC] + [ACE] \\ &= [ABE].\end{aligned}$$

15. Given the figure below with $3BF = 2FC$, $AE = 2EF$, and $[DEF] = 1$.

(a) Find DFC.

(b) Find $\dfrac{AC}{AD}$.

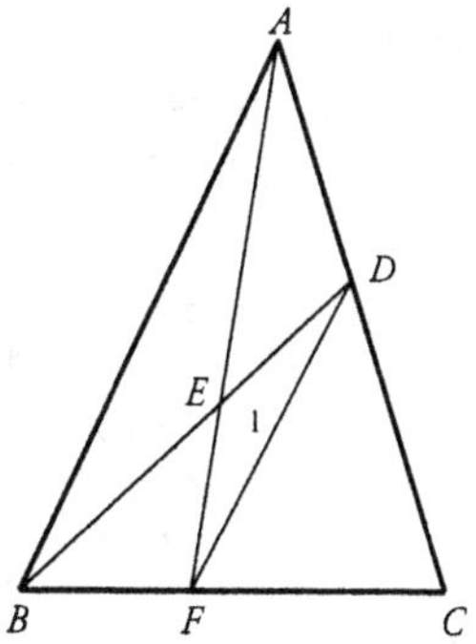

Solution.

(a) Let $[DFC] = y$ and $[BEF] = x$. Since $AE = 2EF$, the areas are as shown in the figure below.

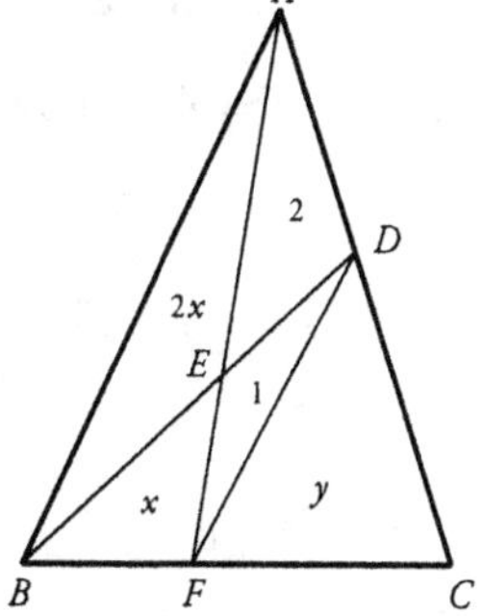

Since $3BF = 2FC$, we have

$$\frac{2}{3} = \frac{BF}{FC} = \frac{[ABF]}{[AFC]} = \frac{3x}{3+y}$$

so that

$$9x = 6 + 2y.$$

Similarly,

$$\frac{2}{3} = \frac{BF}{FC} = \frac{[BDF]}{[DFC]} = \frac{1+x}{y}$$

so that

$$2y = 3 + 3x,$$

and therefore $x = \frac{3}{2}$ and $[DFC] = y = \frac{15}{4}$.

(b) We have

$$\frac{AC}{AD} = \frac{[ABC]}{[ABD]} = \frac{3x+3+y}{2x+2} = \frac{9}{4}.$$

17. Given a rectangle, construct a square having the same area.

Solution. Given a rectangle with sides a and b, construct the segments AB and BC with $AB = a$ and $BC = b$. Next, construct a semicircle with diameter AC and erect a perpendicular to AC at B, hitting the semicircle at D. Finally, construct a square with side BD.

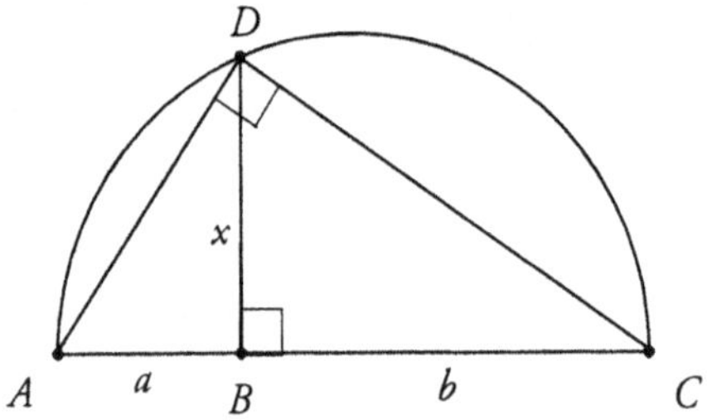

Let $x = BD$. From Thales' Theorem, $\triangle ACD$ is a right triangle. It suffices to show that $x^2 = ab$. We will give two proofs.

First proof. From Pythagoras' Theorem, we have

$$(a+b)^2 = AD^2 + DC^2.$$

Applying Pythagoras' Theorem to $\triangle ADB$ and $\triangle CDB$, we have

$$AD^2 = a^2 + x^2 \quad \text{and} \quad DC^2 = b^2 + x^2.$$

Therefore,

$$(a+b)^2 = a^2 + x^2 + b^2 + x^2 = a^2 + b^2 + 2x^2.$$

That is, $a^2 + 2ab + b^2 = a^2 + b^2 + 2x^2$, so that $x^2 = ab$.

Second proof. In triangles ADC and ABD we have a common angle $\angle A$, and

$$\angle ABD = 90^\circ = \angle ADC,$$

so $\triangle ADC \sim \triangle ABD$ by **AA** similarity condition.
Similarly, $\triangle ADC \sim \triangle DBC$ and so

$$\triangle ABD \sim \triangle DBC$$

and so

$$\frac{BD}{AB} = \frac{BC}{DB}.$$

That is,

$$\frac{x}{a} = \frac{b}{x}$$

so that $x^2 = ab$.

19. P is a point inside an equilateral triangle ABC. Perpendiculars PD, PE, and PF are dropped from P onto BC, CA, and AB, respectively. Prove that

$$[PAF] + [PBD] + [PCE] = [PAE] + [PCD] + [PBF].$$

Solution. In the equilateral triangle ABC, draw lines through P parallel to the sides of the triangle.

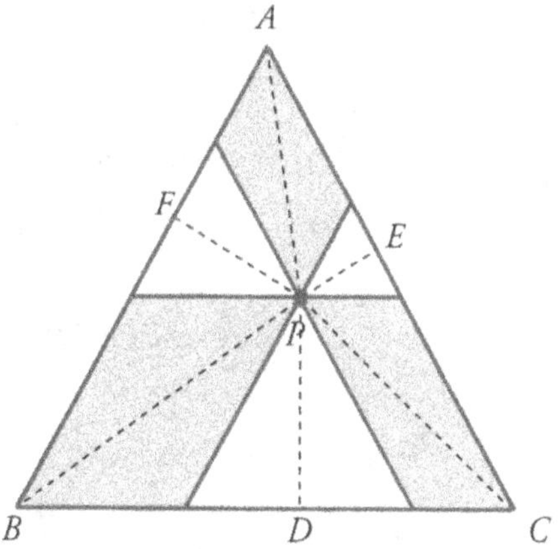

Since the lines PA, PB, and PC bisect the areas of the three parallelograms formed and the areas of the remaining three equilateral triangles are bisected by the altitudes from P, the result follows.

21. A triangle has sides 13, 14, and 15. Find its altitude on the base of length 14.

Solution. Letting h be the altitude on the base 14, by the base altitude formula the area is $\frac{1}{2}14h$.

The semiperimeter is 21, so by Heron's formula, the area is

$$\sqrt{21(21-13)(21-14)(21-15)}.$$

Equating the two expressions for the area and solving for h we get $h = \frac{84}{7}$.

23. $ABDE$, $BCGF$, and $CAHI$ are three squares drawn on the outside of triangle ABC, which has a right angle at C. Prove that

$$GD^2 - HE^2 = 3([BCFG] - [CAHI]).$$

Solution. Draw GD and EH as shown in the figure below.

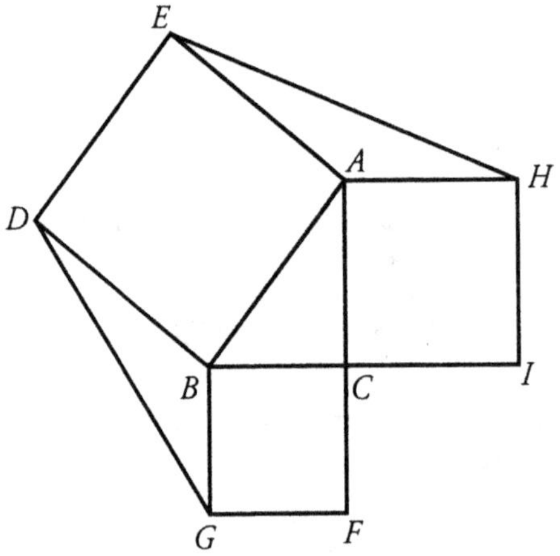

From the Law of Cosines we have

$$GD^2 = BD^2 + BG^2 - 2BD \cdot BG \cos \angle DBG$$
$$EH^2 = AE^2 + AH^2 - 2AE \cdot AH \cos \angle EAH.$$

Now

$$\angle DBC = 180 - \angle B \qquad \text{and} \qquad \angle EAH = 180 - \angle A$$

so that

$$\cos \angle DBG = -\cos B \qquad \text{and} \qquad \cos \angle EAH = -\cos A,$$

and since

$$AE = AB, \quad AH = AC, \quad BD = AB, \quad BG = BC,$$

we have

$$\begin{aligned} GD^2 &= AB^2 + BC^2 + 2BC \cdot AB \cos B \\ &= AB^2 + BC^2 + 2BC^2 = AB^2 + 3BC^2 \\ EH^2 &= AB^2 + AC^2 + 2AC \cdot AB \cos A \\ &= AB^2 + AC^2 + 2AC^2 = AB^2 + 3AC^2. \end{aligned}$$

Therefore,

$$GD^2 - EH^2 = 3(BC^2 - AC^2) = 3([BCFG] - [CAHI]).$$

25. $ABCDE$ is a convex pentagon such that

$$AB = AC, \quad AD = AE, \quad \text{and} \quad \angle CAD = \angle ABE + \angle AEB.$$

If M is the midpoint of BE, prove that $CD = 2AM$.

Solution. In the figure on the following page, let

$$z = \angle CAD, \quad x = \angle ABE, \quad \text{and} \quad y = \angle AEB,$$

so that $z = x + y$.

Since $AB = AC$ and $AD = AE$, then $\triangle ABC$ and $\triangle ADE$ are isosceles. Let AX be the perpendicular bisector of BC and let AY be the perpendicular bisector of DE, then

$$\alpha = \angle BAX = \angle CAX \quad \text{and} \quad \beta = \angle EAY = \angle DAY,$$

as in the figure.

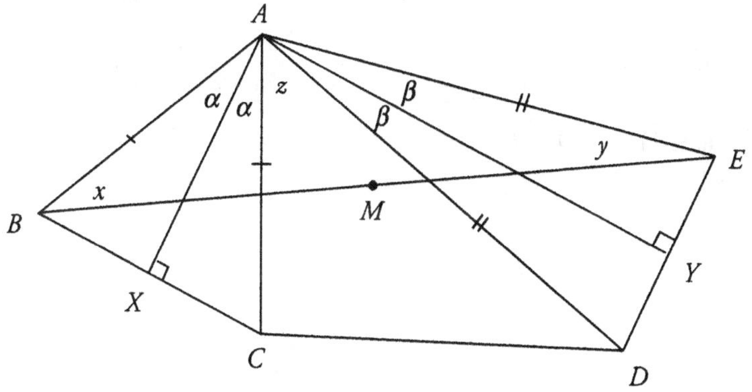

Summing the angles in $\triangle ABE$, and using the fact that $z = x + y$, we have

$$x + 2\alpha + z + 2\beta + y = 180,$$

so that

$$z + \alpha + \beta = 90,$$

and $\angle XAY$ is a right angle.

From the Law of Cosines we have

$$CD^2 = AC^2 + AD^2 - 2AC \cdot AD \cos z$$
$$BE^2 = AB^2 + AE^2 - 2AB \cdot AE \cos(2\alpha + 2\beta + z),$$

and since $2\alpha + 2\beta + z = 180 - z$, then

$$\cos(2\alpha + 2\beta + z) = \cos(180 - z) = -\cos z.$$

Also, since $AB = AC$ and $AD = AE$,

$$BE^2 = AB^2 + AE^2 + 2AB \cdot AE \cos z$$
$$CD^2 = AB^2 + AE^2 - 2AB \cdot AE \cos z.$$

Adding these two equations, we have

$$BE^2 + CD^2 = 2AB^2 + 2AE^2;$$

that is,

$$CD^2 = 2AB^2 + 2AE^2 - BE^2.$$

Now, M is the midpoint of BE in $\triangle ABE$, and from Apollonius' Theorem, we have

$$AB^2 + AE^2 = 2AM^2 + 2BM^2 = 2AM^2 + \frac{1}{2}BE^2,$$

and

$$CD^2 = 2AB^2 + 2AE^2 - BE^2 = 4AM^2;$$

that is, $CD = 2AM$.

CHAPTER 6

MISCELLANEOUS TOPICS

1. Construct a triangle given the three midpoints of its sides.

Solution. Let D, E, and F be the midpoints of the sides BC, AC, and AB, respectively. Draw the triangle $\triangle DEF$ joining the midpoints of the sides of $\triangle ABC$, as shown in the figure below.

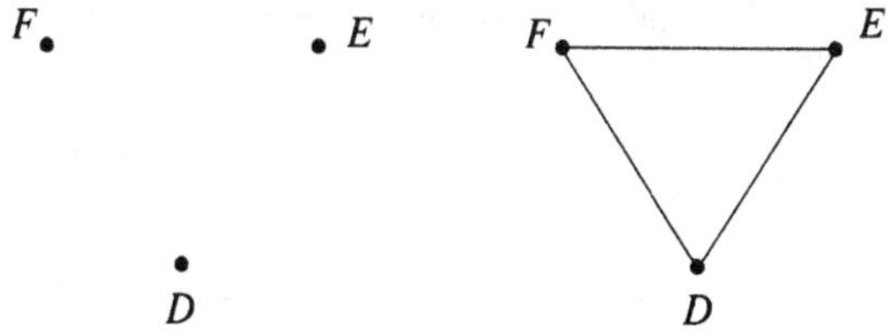

By the Midline Theorem, the point D is the midpoint of a line segment BC parallel to $\overline{EF}$, the point E is the midpoint of a line segment AC parallel to $\overline{DF}$, and the point F is the midpoint of a line segment AB parallel to $\overline{ED}$.

Analysis Figure.

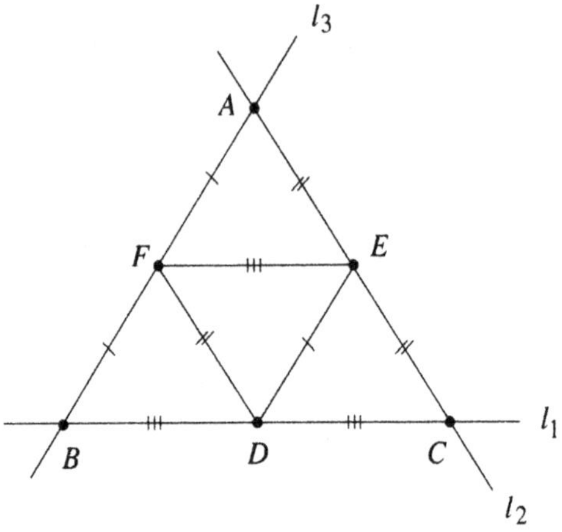

Construction.

(1) Construct the triangle $\triangle DEF$ by joining the midpoints D, E, and F of the sides of $\triangle ABC$.

(2) Construct a line ℓ_1 through D parallel to $\overline{EF}$.

(3) Construct a line ℓ_2 through E parallel to $\overline{DF}$.

(4) Construct a line ℓ_3 through F parallel to $\overline{ED}$. (5) $A = \ell_2 \cap \ell_3$, $B = \ell_1 \cap \ell_3$, $C = \ell_1 \cap \ell_2$.

as in the figure.

Justification.

(a) Since $\ell_2 \parallel DF$ and $\ell_3 \parallel DE$, then $\square AFDE$ is a parallelogram so that $AF = DE$ and $AE = DF$.

(b) Since $\ell_1 \parallel EF$ and $\ell_3 \parallel DE$, then $\square BFED$ is a parallelogram so that $BF = DE$ and $EF = BD$.

(c) Since $\ell_2 \parallel DF$ and $\ell_1 \parallel EF$, then $\square CDFE$ is a parallelogram so that $DF = CE$ and $DC = EF$.

Therefore, D is the midpoint of BC, E is the midpoint of AC, and F is the midpoint of AB.

3. Construct a triangle given the length of one side, the distance from an adjacent vertex to the incenter, and the radius of the incircle.

Solution.

Analysis Figure. We are given $a = BC$, $m = BI$, the distance from the vertex B to the incenter I (we are not given the incenter), and r, the radius of the incircle.

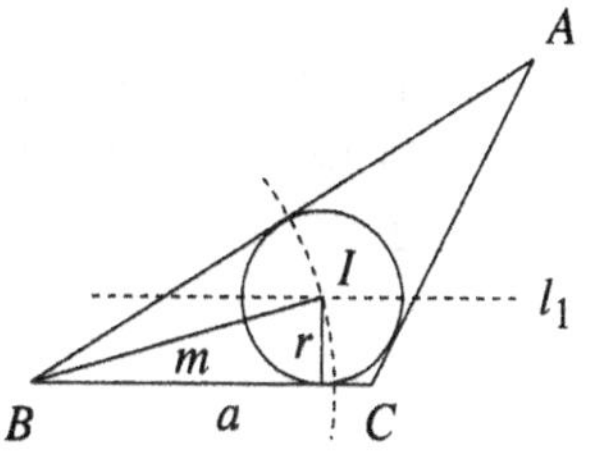

Note that we can find the third vertex A of the triangle if we know the incenter I of the triangle, since we can then construct two tangents ℓ_2 and ℓ_3 to the incircle from the external points B and C, respectively, and $A = \ell_2 \cap \ell_3$.

Key. The incenter I lies on two constructible loci:

(a) A circle centered at B with radius m.

(b) A line parallel to BC and at a perpendicular distance r from BC.

Construction.

(1) Draw the line segment $\overline{BC}$.

(2) Draw the circle $C_1(B, m)$.

(3) Draw the line ℓ_1 parallel to BC and at a perpendicular distance r from BC. The incenter is $I = \ell_1 \cap C(B, m)$.

(4) Draw $C_2(I, r)$.

(5) Construct tangents ℓ_2 and ℓ_3 to C_2 from B and C.

(6) Let $A = \ell_2 \cap \ell_3$.

Then ABC is the desired triangle.

Justification. The incircle is tangent to all three sides of the triangle. The incenter is a perpendicular distance r from each of the sides of the triangle and is at a distance m from the vertex B.

5. Construct a triangle given the measure of one angle, the length of the internal bisector of that angle, and the radius of the incircle.

We give two solutions to this problem.

Solution I.

Analysis Figure. We are given $x = m\angle B$, the length m of the internal bisector of the angle at B, and the radius r of the incircle of the triangle ABC.

Let E be the point of intersection of the angle bisector of $\angle B$ with the opposite side AC. If we can find the incenter I of the triangle, then we can mark off the point E on the ray $\overrightarrow{BI}$ so that $BE = m$.

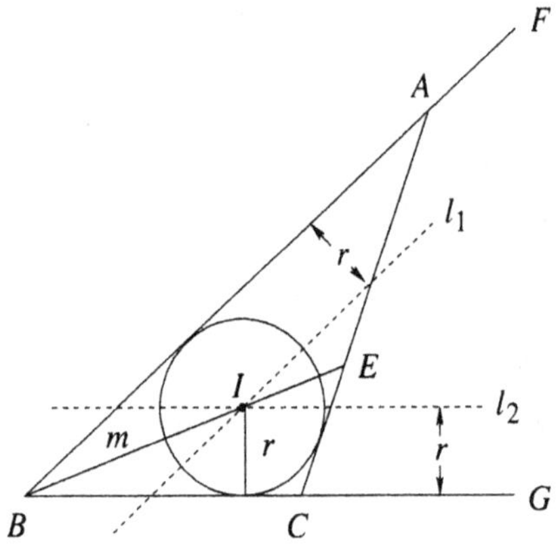

Given the incenter I and the radius r of the incircle, we can construct the incircle $\mathcal{C}(I, r)$ and an external tangent from E to the incircle. The vertices A and C are then the intersection of this tangent line with the arms of the angle at B. Note that if E is on the incircle, then there is a unique solution, whereas if E is not on the incircle, we can draw two tangents from E to the circle.

Key. Note that the incenter I lies on two constructible loci:

(a) A line ℓ_1 parallel to the ray $\overrightarrow{BF}$ and at a perpendicular distance r from the ray.

(b) A line ℓ_2 parallel to the ray $\overrightarrow{BG}$ and at a perpendicular distance r from the ray.

Construction.

(1) Draw the angle $\angle FBG$ of size x.

(2) Draw the line ℓ_1 parallel to $\overrightarrow{BF}$ at distance r.

(3) Draw the line ℓ_2 parallel to $\overrightarrow{BG}$ at distance r.

(4) The incenter is $I = \ell_1 \cap \ell_2$.

(5) Draw the incircle $C(I, r)$.

(6) Draw $\overline{BE}$ through I so that $BE = m$.

(7) Construct a tangent ℓ_3 to $C(I, r)$ from E and let $A = \ell_3 \cap \overrightarrow{BF}$ and $C = \ell_3 \cap \overrightarrow{BG}$.

Then $\triangle ABC$ is the desired triangle.

Justification. The justification is given above, before the construction.

Solution II.

Analysis Figure. If we draw the segment $\overline{BE}$ with length m and copy the angle $x/2$ to both sides of the segment, as in the figure, then the incenter I lies on the segment $\overline{BE}$ and on the line ℓ_1 parallel to $\overline{BF}$ at a perpendicular distance r from $\overline{BF}$.

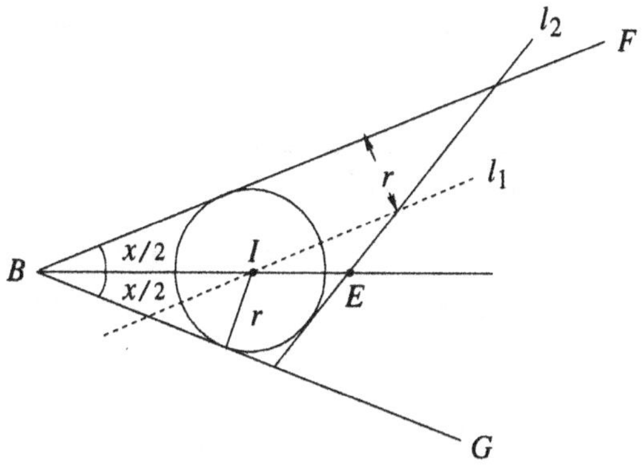

Given the incenter I and the radius r of the incircle, we can construct the incircle $C(I, r)$ and an external tangent from E to the incircle. The vertices A and C are then the intersection of this tangent line with the arms of the angle at B.

Key. Note that the incenter I lies on two constructible loci:

(a) On the segment $\overline{BE}$ that bisects the angle at B.

(b) On a line ℓ parallel to the ray $\overrightarrow{BF}$ and at a perpendicular distance r from the ray.

Construction.

(1) Draw the segment $\overline{BE}$ with $BE = m$.

(2) Copy the angle $x/2$ to both sides of $\overline{BE}$ at B.

(3) Draw the line ℓ_1 parallel to $\overline{BF}$ a distance r units away, hitting $\overline{BE}$ at I.

(4) Draw the incircle $\mathcal{C}(I, r)$.

(5) Draw the tangent ℓ_2 to $\mathcal{C}(I, r)$ from E and let $A = \ell_2 \cap \overrightarrow{BF}$ and $C = \ell_2 \cap \overrightarrow{BG}$.

Then $\triangle ABC$ is the desired triangle.

Justification. The characterization theorem for the internal angle bisector and the definition of the incircle.

7. Given segments of length 1, a, and b, explain how to solve the following geometrically:

$$x^2 = a + b^2.$$

Explain how to construct the segment of length x.

Solution. We wish to construct a segment of length $\sqrt{a + b^2}$. Here are the construction steps.

(a) Referring to Example 6.2.2, construct a segment PQ of length pq where $p = q = b$; that is, $PQ = b^2$.

(b) Extend this segment to obtain a segment PS of length $a + b^2$.

(c) Construct the segment of length $\sqrt{a + b^2}$ using Example 6.2.3. (In Example 6.2.3, let $AD = 1$ and $DB = a + b^2$.)

9. Can we construct an angle of $2°$?

Solution. This is the central angle for an n-gon where

$$n = \frac{360}{2} = 180.$$

The prime factorization of 180 is

$$180 = 2^2 \cdot 3^3 \cdot 5,$$

so an angle of 2o is not constructible by Gauss' Theorem.

11. Prove Miquel's Theorem for the case where two of the circles are tangent. That is, given $\triangle ABC$ with menelaus points X, Y, and Z, as shown, where the circumcircles of $\triangle AXZ$ and $\triangle BXY$ are tangent at X, show that the quadrilateral $XYCZ$ is cyclic.

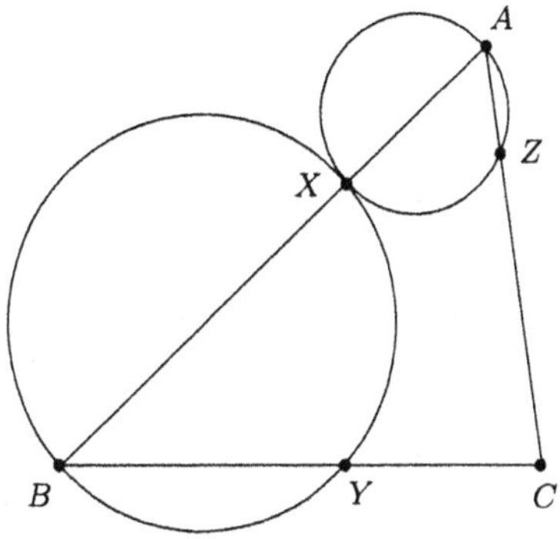

Hint. Tangent circles at X have the same tangent line to both circles at X.

Solution. Draw the common tangent XP to the two circumcircles and let $\angle PXZ = r$ and $\angle PXY = s$.

From the corollary to Thales' Theorem, the angle between a chord and the tangent is equal to the angle subtended by the chord at the circumference so that

$$\angle XAZ = \angle PXZ = r \qquad \text{and} \qquad \angle XBY = \angle PXY = s.$$

The angles are as shown in the figure below.

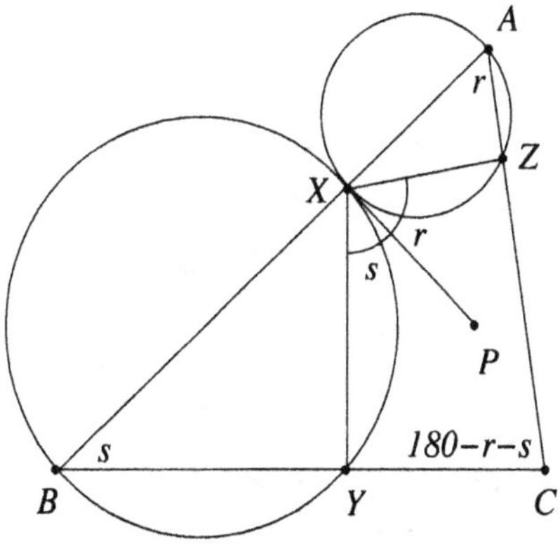

Since opposite angles in the simple quadrilateral $XYCZ$ are supplementary, $\square XYCZ$ is cyclic.

13. Given a diameter $\overline{AB}$ of a circle and a point P as shown, construct a perpendicular from P to $\overleftrightarrow{AB}$, *with a straightedge alone.*

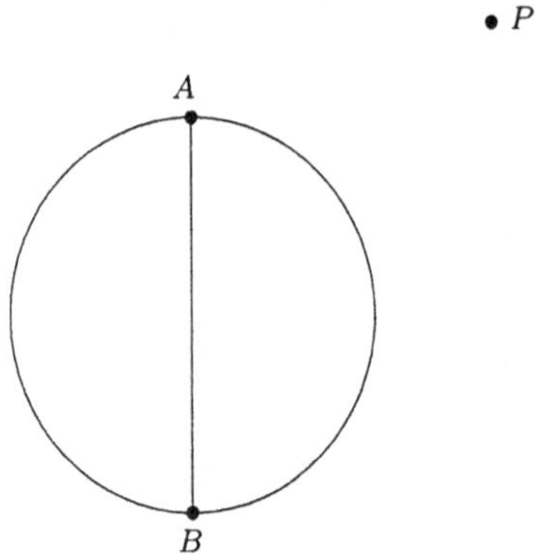

Solution.

Analysis Figure. Join P to A and let the line segment $\overline{PA}$ hit the circle at the point D. Next, join P to B, and let the line segment $\overline{PB}$ hit the circle at the point C. Finally, let $\overline{CA}$ and $\overline{BD}$ intersect at the point Q.

Since triangles ADB and ACB are both inscribed in a semicircle, they are right triangles with the right angles at D and C, respectively.

Note that the altitudes $\overline{PD}$ and $\overline{QC}$ of $\triangle PQB$ intersect at the point A so that A is the orthocenter and BM must be the third altitude.

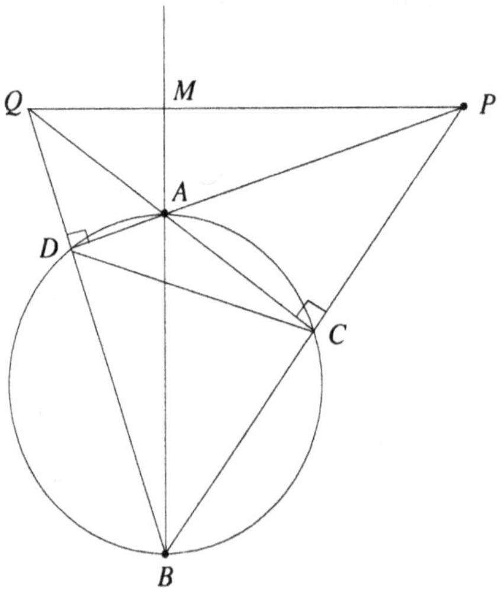

Construction.

(a) Join P to A and let the line segment $\overline{PA}$ hit the circle at the point D.

(b) Join P to B and let the line segment $\overline{PB}$ hit the circle at the point C.

(c) Draw $\overline{CA}$ and $\overline{BD}$ intersecting at the point Q.

The line segment $\overline{PQ}$ is perpendicular to $\overline{AB}$.

Justification. As above.

15. Construct $\triangle ABC$ given a line segment $\overline{XY}$ and two adjacent angles, as in the figure, where the length of the perimeter is XY, $\alpha = \angle B$, and $\beta = \angle C$.

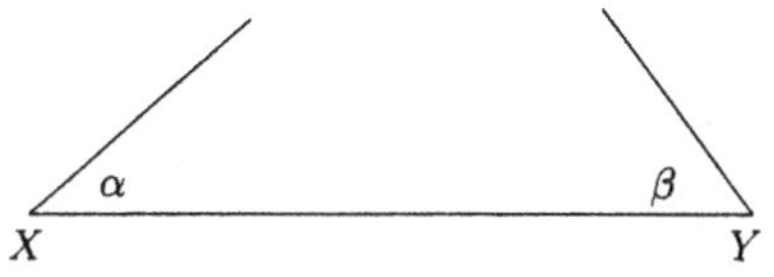

Solution. Let the angle bisectors of the angles at X and Y intersect at the point A and let l and m be the perpendicular bisectors of the sides AX and AY, respectively, hitting AX at M and AY at N, the midpoints of the sides.

Finally, let B and C be the points of intersection of l and m with the segment $\overline{XY}$, as shown below, then $\triangle ABC$ is the desired triangle.

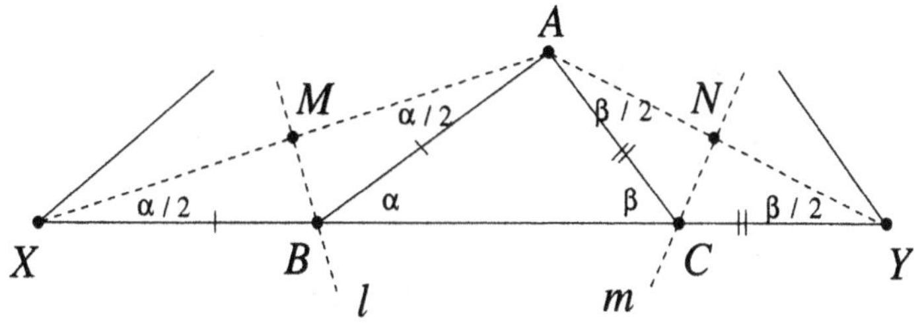

Since l is the perpendicular bisector of the side AX, B is equidistant from the points A and X. Similarly, since m is the perpendicular bisector of the side AY, C is equidistant from the points A and Y so that

$$BX = AB \qquad \text{and} \qquad CY = AC.$$

Also,

$$MX = MA \qquad \text{and} \qquad NA = NY,$$

and by the SAS congruency theorem, we have

$$\triangle BMX \equiv \triangle BMA \qquad \text{and} \qquad \triangle CNY \equiv \triangle CNA.$$

From the Exterior Angle Theorem, we have

$$\angle B = \frac{\alpha}{2} + \frac{\alpha}{2} = \alpha$$

and

$$\angle C = \frac{\beta}{2} + \frac{\beta}{2} = \beta,$$

and also

$$AB + BC + CA = XB + BC + CY = XY.$$

Therefore, $\triangle ABC$ has $\angle B = \alpha$, $\angle C = \beta$, and perimeter XY.

17. Construct a triangle given the foot F of an altitude, the circumcenter S, and the center N of the 9-point circle.

Solution.

Analysis Figure.

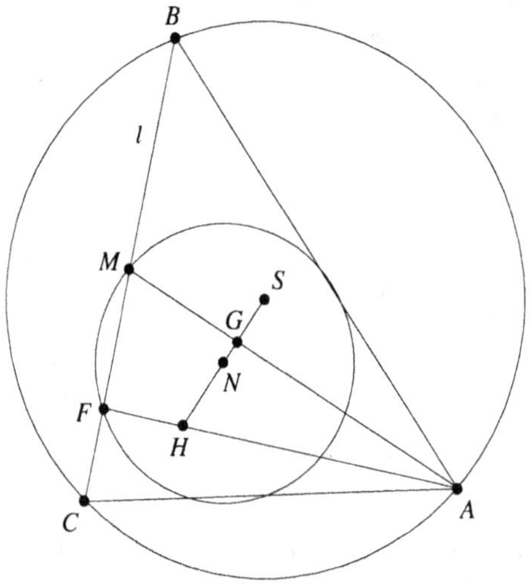

Construction.

(a) Draw the line SN and extend it past N to H so that $SN = NH$. This makes H the orthocenter and N the midpoint of SH. Also, draw the line SG so that $GH = 2SG$, which makes G the centroid.

(b) Draw the line ℓ perpendicular to FH through F.

(c) Draw the circle $C_1(N, NF)$, hitting the line ℓ at F and M.

(d) Draw the line joining M and G, hitting the line FH at A.

(e) Draw the circle $C_2(S, AS)$, hitting the line through F and M at B and C.

The triangle ABC is the desired triangle.

Justification. Since N is the midpoint of the segment joining the circumcenter with the orthocenter, the first step makes H the orthocenter, so that the line FH is the altitude from A, and the second step makes G the centroid.

The midpoint M of the segment BC is on the 9-point circle $\mathcal{C}_1(N, NF)$ and the median from A goes through M and G, so the vertex A is the intersection of the lines MG and FH.

Since the vertices B and C are on the line perpendicular to AH passing through F, they are the intersection points of the circumcircle $\mathcal{C}_2(S, AS)$ with this line.

19. Show that the incenter of a triangle is the Nagel point of its medial triangle.

Solution. Given triangle ABC, the *medial triangle* of $\triangle ABC$ is the triangle DEF, with vertices D, E, and F the midpoints of the sides $\overline{BC}$, $\overline{AC}$, and $\overline{AB}$, respectively, as in the figure below.

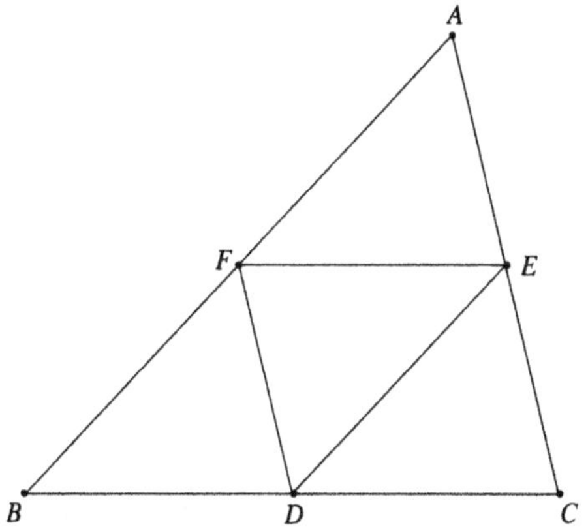

From the Midline Theorem, we have

$$\overline{FE} \parallel \overline{BC} \quad \text{and} \quad EF = \frac{1}{2}BC$$

$$\overline{FD} \parallel \overline{AC} \quad \text{and} \quad DF = \frac{1}{2}AC$$

$$\overline{ED} \parallel \overline{AB} \quad \text{and} \quad DE = \frac{1}{2}AB.$$

Let R be the Nagel point of $\triangle DEF$. If $s = (DE + EF + FD)/2$ is the semi-perimeter, then $\overline{DR}$ intersects the side $\overline{EF}$ at L and $\overline{ER}$ intersects the side $\overline{DF}$ at M, where

$$EL = s - ED \quad \text{and} \quad DM = s - ED,$$

so that $EL = DM$.

Now we extend $\overline{ER}$ so that it intersects $\overline{BC}$ at P, then the angles are as shown in the figure on the following page, and

$$\triangle PRD \sim \triangle ERL,$$

since $\overline{BC} \parallel \overline{EF}$, while

$$\triangle DPM \sim \triangle CPE,$$

since $\overline{FD} \parallel \overline{CE}$.

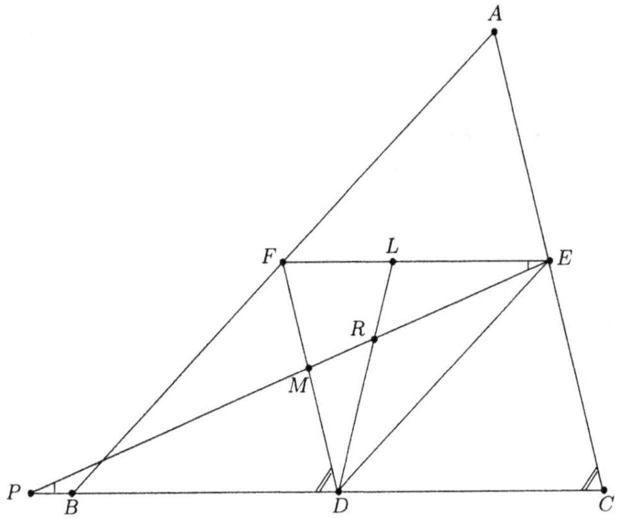

Since $\triangle PRD \sim \triangle ERL$ and $EL = DM$, we have

$$\frac{PR}{RE} = \frac{DP}{EL} = \frac{DP}{DM},$$

and, since $\triangle DPM \sim \triangle CPE$, we have

$$\frac{DP}{DM} = \frac{CP}{CE}.$$

Therefore,

$$\frac{PR}{RE} = \frac{CP}{CE},$$

and by the converse of the Angle Bisector Theorem applied to $\triangle ECP$, this implies that R is on the angle bisector of $\angle ACB$.

A similar argument shows that R is also on the angle bisector of $\angle ABC$ and on the angle bisector of $\angle BAC$.

Therefore, the Nagel point R of the medial triangle DEF is also the incenter of the triangle ABC.

PART II

TRANSFORMATIONAL GEOMETRY

CHAPTER 7

EUCLIDEAN TRANSFORMATIONS

1. A swimming pool is posting notices to inform visitors about their Monday closure. All the notices are identical except for the one by the diving board. It reads: "NOW NO SWIMS ON MON." What could be the reason for this?

   Solution. The divers are upside down most of the time. This notice reads the same upside down as right side up.

3. The diagram below shows five symbols in a sequence. What could the sixth symbol be?

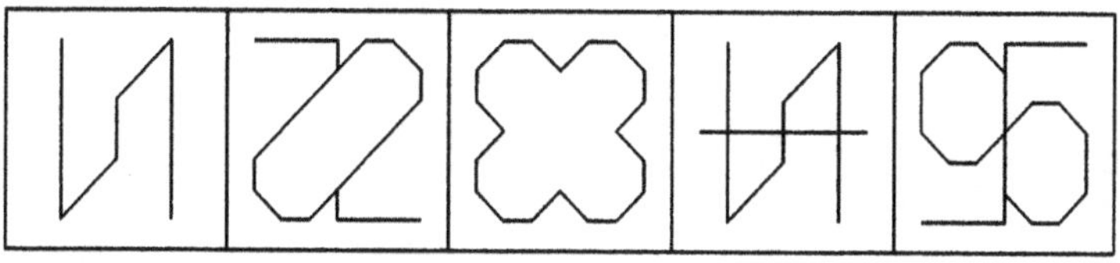

Solution. The common characteristic of these symbols is that they all have a center of symmetry. When we take only the right half of each symbol, we get schematic representations of the numbers 1, 2, 3, 4, and 5. Following this pattern, the sixth symbol should be a schematic representation of the number 6 duplicated by a half-turn, as shown in the figure below.

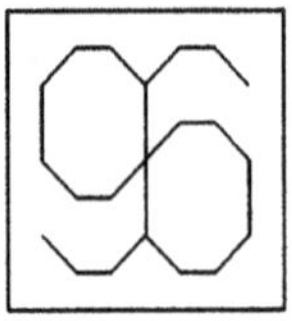

5. Dissect each of the figures below into two congruent pieces.

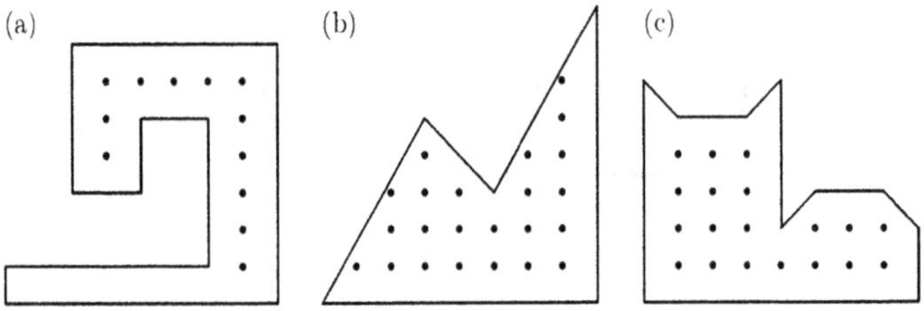

Solution. The solutions are given in the figures below.

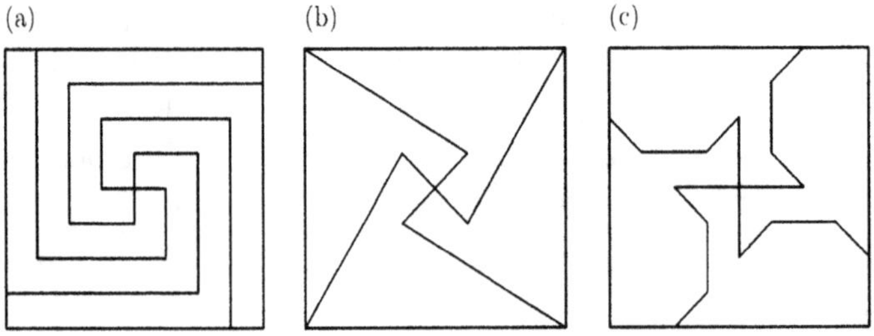

7. Dissect each of the figures below into two congruent pieces.

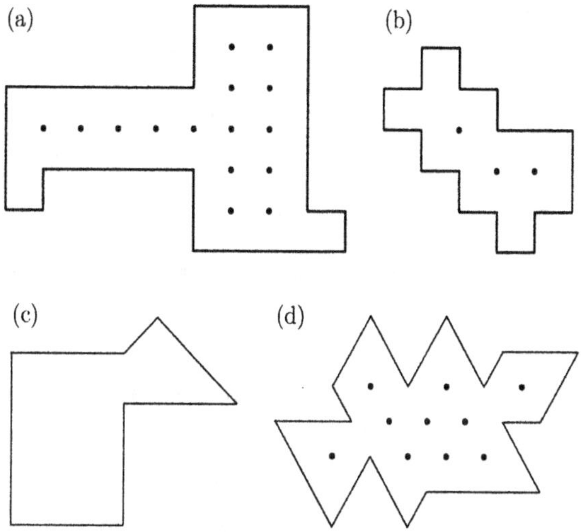

Solution. The solutions are given in the figures below.

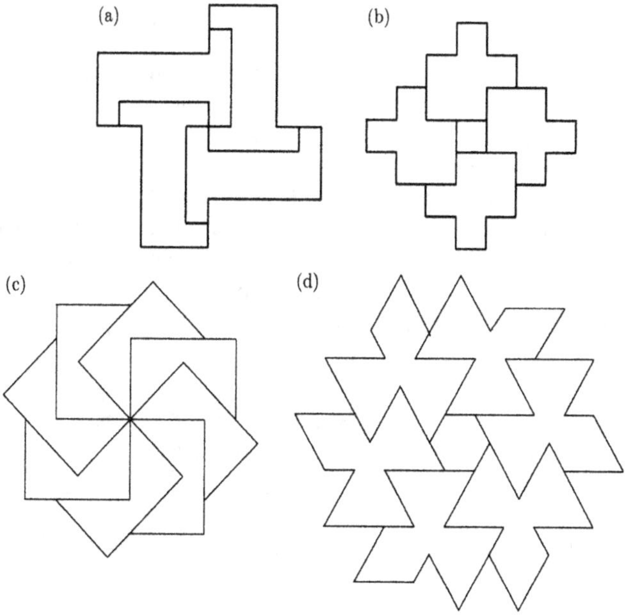

9. Dissect each of the figures in the diagram below into two congruent pieces.

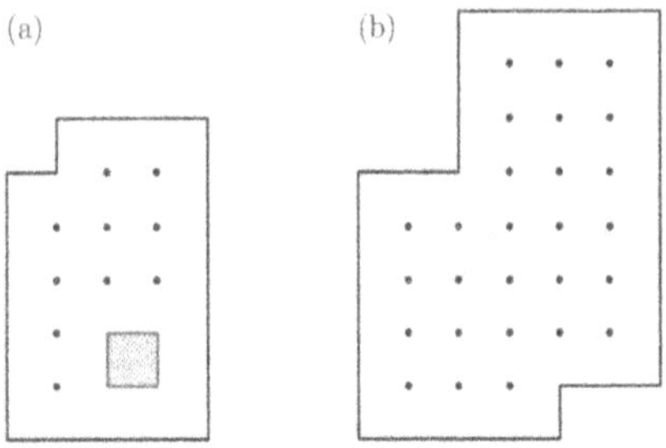

Solution. The solutions are shown in the figures below and were obtained using the method in Example 7.3.2.

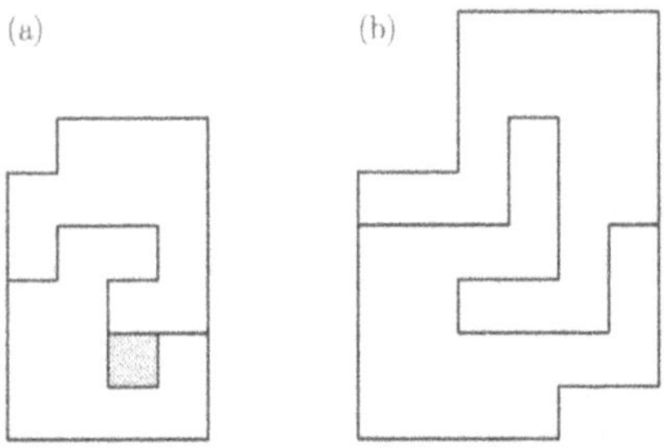

11. Dissect a circle into a number of congruent pieces such that at least one piece does not include the center of the circle, either in its interior or on its boundary. Find two solutions.

Solution. The figure below on the left shows a dissection of a circle into six congruent pieces, each of which has the center of the circle on its boundary. The figures below in the middle and on the right show two ways that each piece can be dissected in the same manner into two congruent pieces. Now six of these twelve pieces have the desired property.

13. Describe how to inscribe a parallelogram with center P in the quadrilateral $ABCD$ shown below.

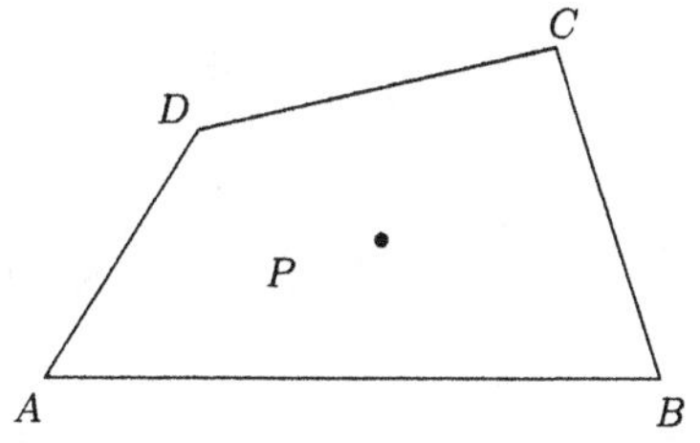

Solution. The diagonals of the desired parallelogram will bisect each other at the center P, so a rotation of 180° about P will map the parallelogram onto itself. Thus, we rotate the quadrilateral $ABCD$ about P through 180° to obtain the quadrilateral $A'B'C'D'$, as in the figure below.

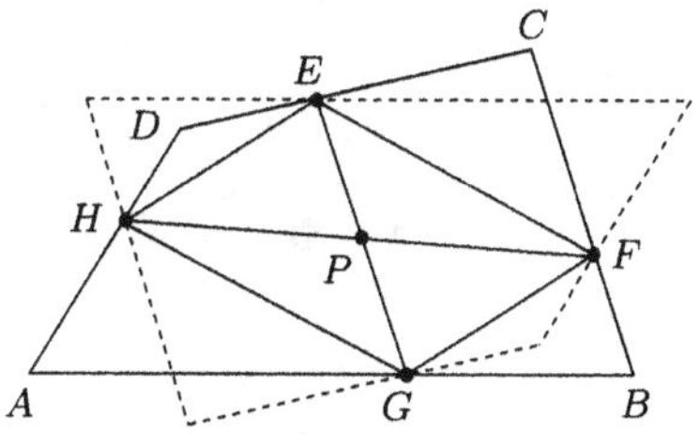

The vertices of the desired parallelogram must be on both quadrilaterals, so we can find the parallelogram among the points where the two quadrilaterals intersect.

Note that in this case there are other solutions in addition to the one shown above.

15. Given three concentric circles, construct an equilateral triangle with one vertex on each circle.

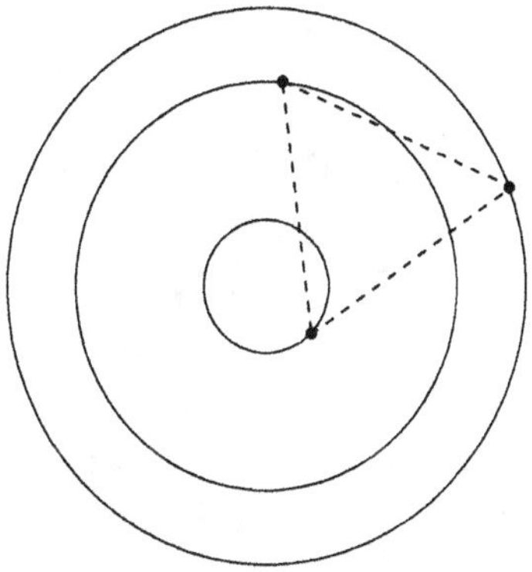

Solution. Let ω_1, ω_2, and ω_3 be the circles in increasing order of size. Let P_1 be an arbitrary point on ω_1, as in the figure below.

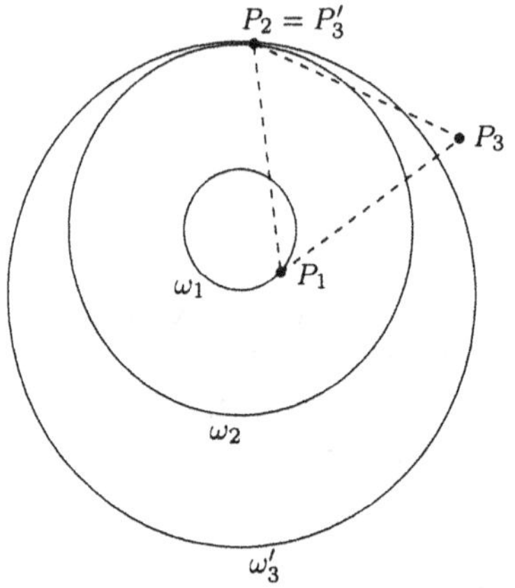

Perform a 60° rotation about P_1 and let ω_3' be the image of ω_3, intersecting ω_2 at the point P_2.

It is now easy to complete the equilateral triangle $P_1P_2P_3$, and P_3 will lie on ω_3 because its image P_3' under the rotation is P_2. In other words, the inverse transformation of the rotation takes P_2 to P_3.

17. Describe how to find all points X on circle $\mathcal{D}_1$ and Y on circle $\mathcal{D}_2$ so that X, P, and Y are collinear and $XP = YP$.

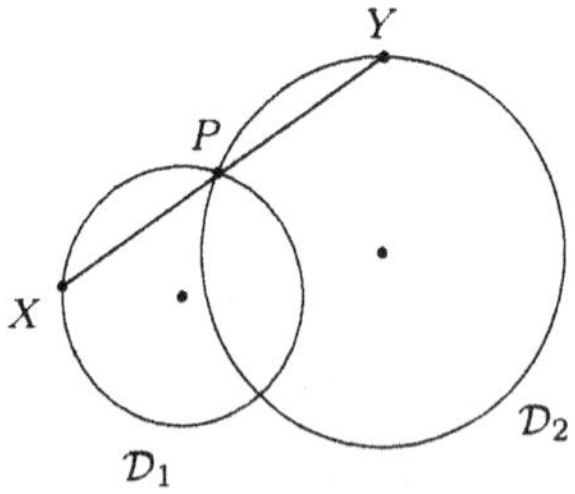

Solution. This problem is solved by applying a rotation $R_{P,180^\circ}$ of 180° about P to circle $\mathcal{D}_1$, obtaining the image $\mathcal{D}_1'$, as shown in the figure on the following page.

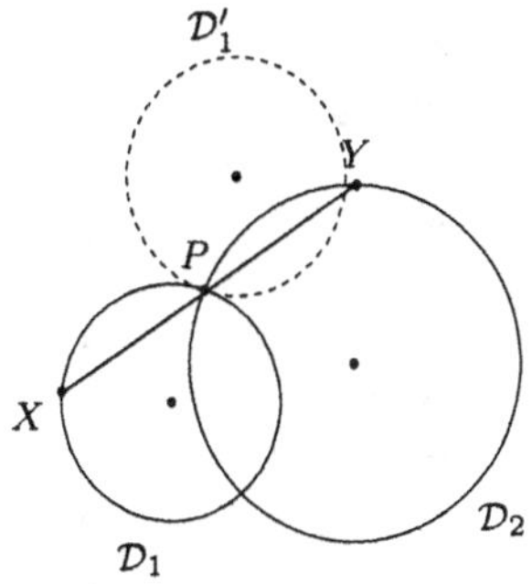

Circles $\mathcal{D}_1'$ and $\mathcal{D}_2$ intersect at points P and Y, and the line from Y through P intersects $\mathcal{C}_1$ at X. Since Y is the image of X under $R_{P,180^\circ}$, then $XP = PY$ and we are done.

19. Two facing mirrors OX and OY form an angle at O. A light ray $ABCD$ reflects off each mirror once, as shown in the figure on the right. If the ray's final direction is opposite to its initial direction, what is the measure of the angle between the mirrors?

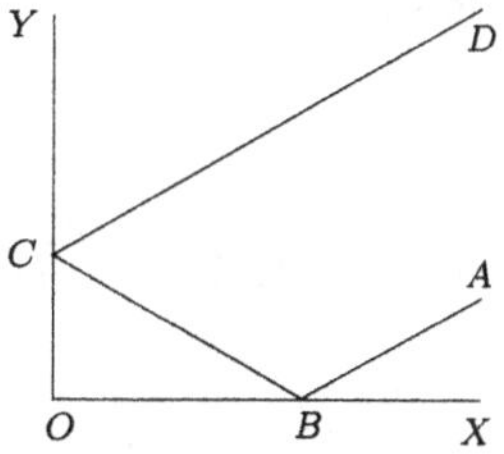

Solution. Draw the line OP parallel to BA and CD, then

$$\angle YOP = \angle YCD = \angle OCB$$

and

$$\angle XOP = \angle XBA = \angle OBC.$$

Now, $\angle XOY = 90^\circ$ since, as can be seen from the figure below, we have

$$\begin{aligned} 180^\circ &= \angle BOC + \angle OCB + \angle OBC \\ &= \angle BOC + \angle YOP + \angle XOP \\ &= 2\angle XOY. \end{aligned}$$

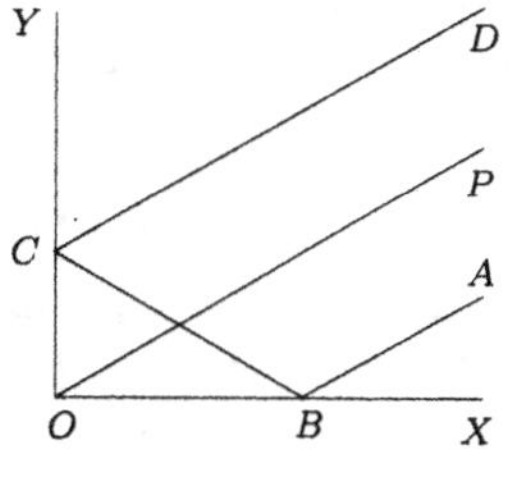

21. Two facing mirrors OX and OU form an angle at O. A light ray $ABCDEFGH$ reflects off each mirror thrice, as shown in the diagram below (G and H not shown). If the ray's final direction is opposite to its initial direction, what is the measure of the angle between the mirrors?

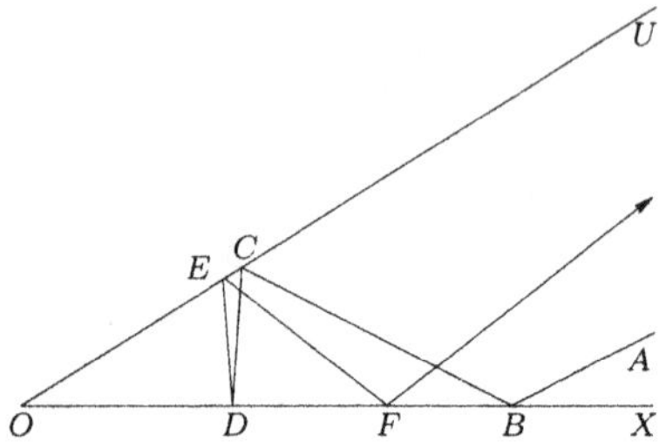

Solution. Reflect OX in the line OU to get OV, mapping D and F into D' and F', respectively.

Note that

$$\angle D'CO = \angle DCO = \angle BCU,$$

and hence C lies on BD'. Similarly, E lies on FD' and DF'.

Reflect OU in the line OV to obtain OY, mapping E into E'. As before, D' lies on CE'. By Problem 7.19, $\angle XOY = 90°$ and it follows that $\angle XOU = 30°$.

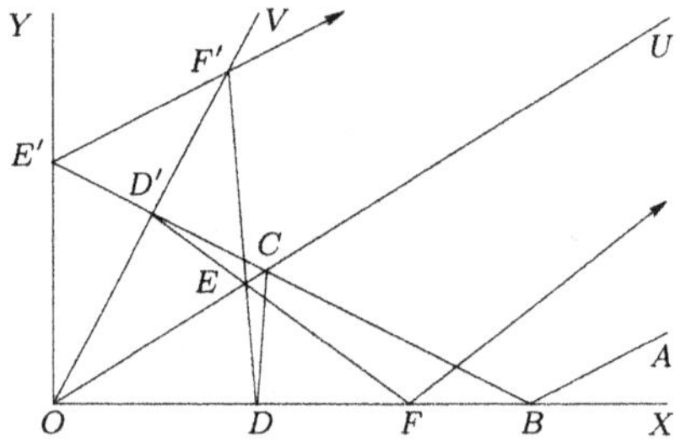

23. A woodsman's hut is in the interior of a peninsula which has the form of an acute angle. The woodsman must leave his hut, walk to one shore of the peninsula, then to the other shore, then return home. How should he choose the shortest such path?

Solution. As in the figure on the following page, reflect the hut A in both shores to obtain points B and C. Let D and E be the points of intersection of BC with the shores, as shown in the figure.

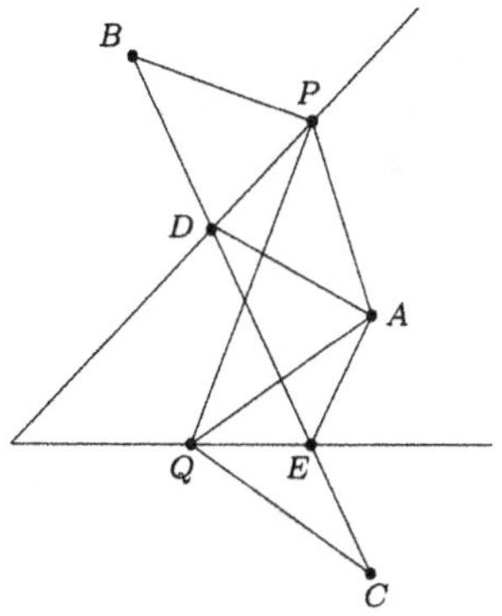

If P is the point at which the woodsman reaches the first shore and Q is the point at which he reaches the second shore, then the route $AP + PQ + QA$ is equal in length to the route $BP + PQ + QC$, since we are simply reflecting the segments AP and AQ in the shores.

However,

$$BP + PQ + QC \geq BC = BD + DE + EC,$$

and it follows that $AD + DE + EA$ is the optimal path.

25. The figure below represents one hole on a mini golf course. The ball B is 5 units from the west wall and 16 units from the south wall. The hole H is 4 units from the east wall and 8 units from the north wall. What is the length of the shortest path for the ball to go into the hole in one stroke, bouncing off a wall only once?

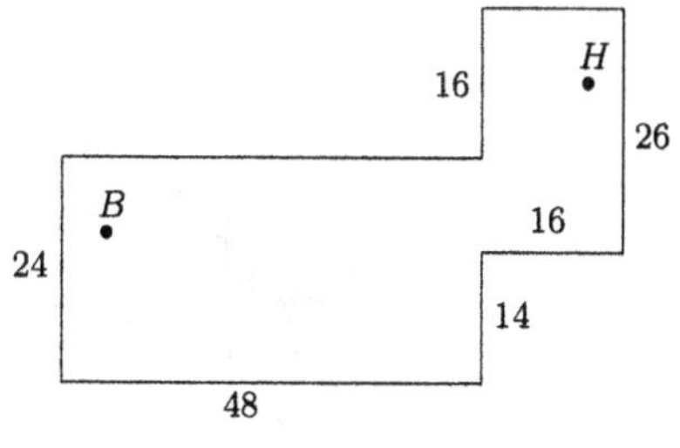

Solution. Reflect B across the south wall to B'. The straight line $B'H$ intersects the south wall at P, and the broken path $B - P - H$ stays within the walls. Complete the right triangle $B'HQ$, as shown in the figure on the following page.

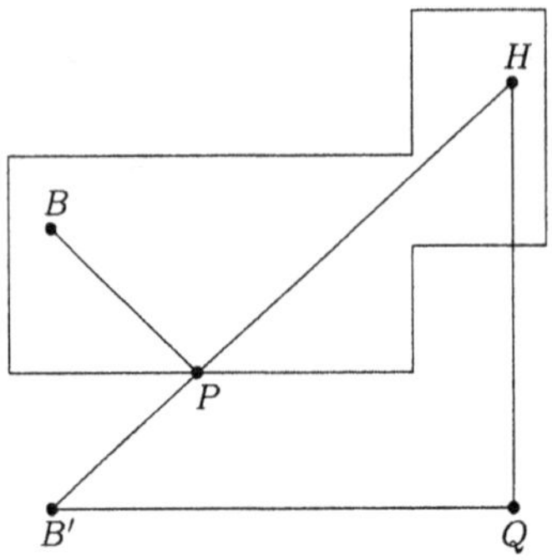

Now, the distance between the east and west walls is 64 units, and the distance between the north and south walls is 40 units. Hence,

$$QB' = 64 - 5 - 4 = 55$$

and

$$QH = 406 - 8 + 16 = 48,$$

and it follows that

$$BP + PH = B'H = \sqrt{55^2 + 48^2} = 73.$$

27. Two circles intersect at two points. Through one of these points P, construct a straight line intersecting the circles again at A and B such that $PA = PB$.

Solution. Let the circles be ω_1 and ω_2. Perform a halfturn about P and let the image ω_2' of ω_2 intersect ω_1 at a point A different from P, as in the figure below.

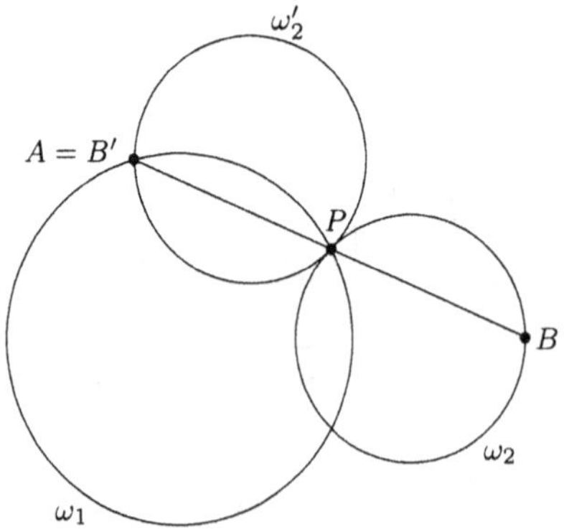

Join AP and extend it to intersect ω_2 again at B, then A coincides with the image B' of B under the halfturn so that $PA = PB$.

29. Two disjoint circles are on the same side of a straight line ℓ. Construct a tangent to each circle so that they intersect on ℓ and make equal angles with ℓ. Find all solutions.

Solution. Let the circles be ω_1 and ω_2 and perform a reflection across ℓ, as in the figure below.

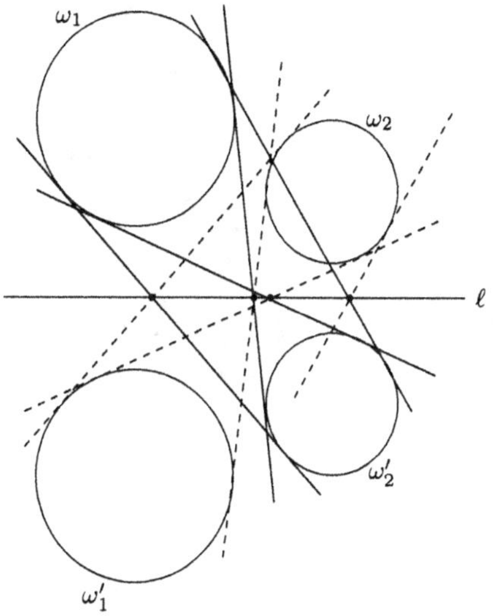

The image ω_2' of ω_2 has four common tangents (solid lines) with ω_1, each of which yields a solution (dashed lines) to the problem, as shown in the figure above.

CHAPTER 8

THE ALGEBRA OF ISOMETRIES

1. Let S and T be two involutive transformations of the plane.

   (a) Prove that ST is involutive if and only if $ST = TS$.

   (b) Assume that S, T, and I are distinct transformations, where I is the identity, such that
   $$ST = TS = X.$$
   Let $\Gamma = \{I, S, T, X\}$. Prove that Γ is a commutative subgroup of $\mathcal{G}$, the group of all transformations on the plane, by constructing the multiplication table.

Solution.

(a) If S and T are involutions, then
$$I = (ST)^2 = STST$$
if and only if
$$S^2TST^2 = SIT = ST,$$

Solutions Manual to Accompany Classical Geometry: Euclidean, Transformational, Inversive, and Projective, First Edition. By I. E. Leonard, J. E. Lewis, A. C. F. Liu, G. W. Tokarsky.

that is, if and only if

$$ITSI = ST,$$

that is, if and only if

$$TS = ST.$$

(b) Note that $X^2 = (ST)^2 = I$ from part (a), so that X is an involution. Therefore,

$$I^{-1} = I, \qquad S^{-1} = S, \qquad T^{-1} = T, \qquad \text{and} \qquad X^{-1} = X,$$

so Γ is closed under taking inverses.

Also,

$$ST = TS = X, \quad SX = S^2T = T = TS^2 = XS,$$
$$TX = T^2S = S = ST^2 = XT$$

so Γ is closed under multiplication and multiplication is commutative. Therefore, Γ is a commutative subgroup of $\mathcal{G}$.

3. Let T be an isometry of the plane. Show that if P and Q are fixed points of T, then every point X on the line through P and Q is a fixed point of T.

Solution. If $T(P) = P$ and $T(Q) = Q$, then $T(P)$ and $T(Q)$ are on the line $T(\ell_{PQ})$; that is, P and Q are on the line $T(\ell_{PQ})$. Since T maps lines onto lines, $T(\ell_{PQ}) = \ell_{PQ}$.

Therefore, if X is a point on the line ℓ_{PQ}, then $T(X)$ is on the line ℓ_{PQ}, and since T is distance preserving, then

$$d(T(X), P) = d(T(X), T(P)) = d(X, P)$$

and

$$d(T(X), Q) = d(T(X), T(Q)) = d(X, Q).$$

Since P, Q, X, and $T(X)$ are all on the same line ℓ_{PQ}, and the isometry T preserves betweenness, then $T(X) = X$.

5. Let S and T be isometries and let A, B, and C be three noncollinear points for which

$$S(A) = T(A), \qquad S(B) = T(B), \qquad \text{and} \qquad S(C) = T(C).$$

Show that $S = T$.

Solution. If we let $\gamma = ST^{-1}$, then γ is an isometry and

$$\gamma(A) = A, \qquad \gamma(B) = B, \qquad \text{and} \qquad \gamma(C) = C,$$

and from the previous problem, since A, B, and C are noncollinear, then

$$\gamma = ST^{-1} = I,$$

that is, $S = T$.

7. Let D, E, and F be the midpoints of the sides BC, AC, and AB, respectively, of $\triangle ABC$ and let T be the transformation of the plane that is the product of the three halfturns

$$T = H_E \, H_D \, H_F.$$

Show that the vertex A of $\triangle ABC$ is a fixed point of T; that is, that $T(A) = A$.

Hint. Draw the picture.

Solution. In the figure below, note that

$$H_F(A) = B, \qquad H_D(B) = C, \qquad \text{and} \qquad H_E(C) = A.$$

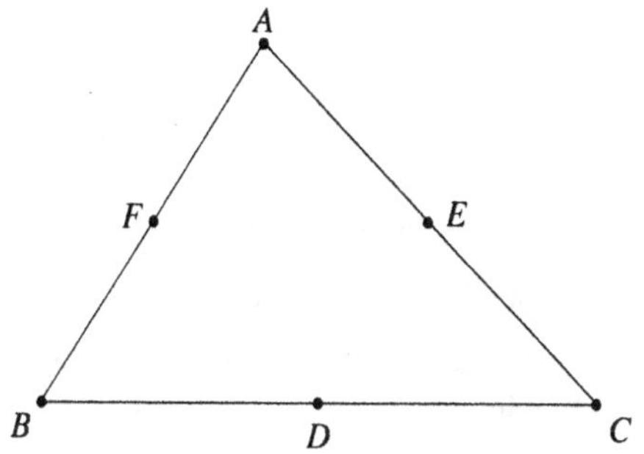

Therefore,

$$T(A) = H_E H_D H_F(A) = H_E H_D(B) = H_E(C) = A,$$

and A is a fixed point of the transformation T.

9. Show that if a circle is invariant under the isometry T, then its center

Solution. Let P be any point on the circle $\mathcal{C}$ with center O and radius a. If $\mathcal{C}$ is invariant under the isometry T, then $T(P)$ is on $\mathcal{C}$ for every P on $\mathcal{C}$. Therefore,

$$d(O, P) = d(T(O), T(P)) = a$$

for every $P \in \mathcal{C}$.

Since T maps the circle $\mathcal{C}$ onto the circle, then given any point Q on the circle, there exists a point P on the circle such that $Q = T(P)$, so that

$$d(T(O), Q) = d(T(O), T(P)) = a,$$

that is,

$$d(T(O), Q) = a$$

for each Q on the circle $\mathcal{C}$. This means that each point on the circle is equidistant from the point $T(O)$, that is, $T(O) = O$, the center of the circle. Thus, $T(O) = O$, and the center O is a fixed point of the isometry T.

11. Let T be an isometry that is an involution and has **exactly** one fixed point O in the plane. Show that T is the halfturn H_O about the point O.

    Solution. First we show that if T is an isometry that is an involution, then for any P in the plane, the midpoint of the line segment joining P and $T(P)$ is a fixed point of T.

    If P is a fixed point of T, then we are done. If P is not a fixed point of T, let M be the midpoint of the line segment from P to $T(P)$, then

    $$d(M, P) = d(M, T(P))$$

    so that

    $$d(T(M), T(P)) = d(T(M), T^2(P)) = d(T(M), P).$$

    Thus, $T(M)$ is on the perpendicular bisector of the line segment from P to $T(P)$.

    Since T maps the line $\ell = \ell_{PT(P)}$ onto itself, then $T(M)$ is also on the line ℓ and

    $$d(P, T(M)) = d(T(P), T(M))$$

    so that $T(M) = M$ and M is a fixed point of T.

    Now suppose that T is an involutive isometry that has a *unique* fixed point O. Let P be any point in the plane with $P \neq O$ and let $P' = T(P)$, then

    $$T(P) = P' \neq P,$$

    and the midpoint of the line segment joining P and $T(P)$ is a fixed point of T. Since T has a unique fixed point, this midpoint is O.

    Therefore, for each point P, the point O is the midpoint of the segment $[P, T(P)]$ and so $T = H_O$.

13. Let T be an isometry of the plane and let ℓ be the perpendicular bisector of the segment $\overline{AB}$. Prove that $T(\ell)$ is the perpendicular bisector of the segment $\overline{T(A)T(B)}$.

    Solution. Let M be the midpoint of the segment $[A, B]$, then

    $$d(M, A) = d(M, B).$$

    Since T is an isometry,

    $$d(T(M), T(A)) = d(M, A) = d(M, B) = d(T(M), T(B))$$

    so that $T(M)$ is on the perpendicular bisector of $[T(A), T(B)]$.

Similarly, if P is any other point on the perpendicular bisector of the segment $[AB]$, then

$$d(T(A), T(P)) = d(A, P) = d(B, P) = d(T(B), T(P))$$

so that $T(P)$ is also on the perpendicular bisector of $[T(A), T(B)]$. Since ℓ is the line passing through M and P, $T(\ell)$ is the line passing through $T(M)$ and $T(P)$; that is, $T(\ell)$ is the perpendicular bisector of the segment $[T(A), T(B)]$.

15. Show that if m and n are perpendicular lines that intersect at a point P in the plane, then

$$R_n\, R_m = H_P.$$

Solution. Let A be an arbitrary point in the plane. We will show that

$$R_n(R_m(A)) = H_P(A).$$

We let $A' = R_m(A)$ and $A'' = R_n(A')$, as in the figure.

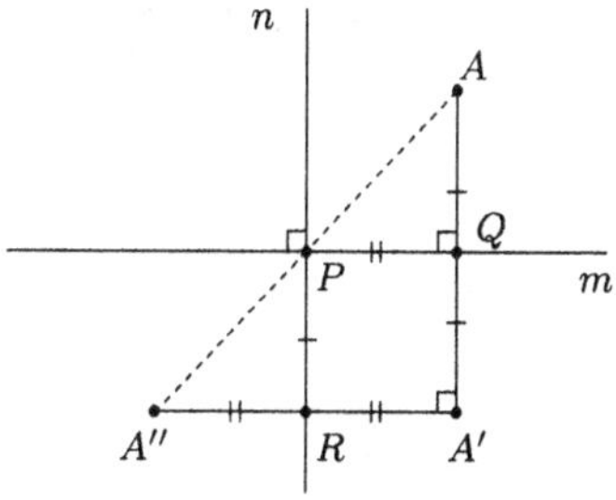

Since $\triangle AQP$ and $\triangle PRA''$ are congruent by the **SAS** congruency theorem, then $\angle APQ = \angle PA''R$ so that A, P, and A'' are collinear. Also,

$$AP = A''P$$

so that $A'' = H_P(A)$.

Therefore,

$$H_P(A) = A'' = R_n(A') = R_n(R_m(A))$$

for all A in the plane, and hence $R_n\, R_m = H_P$.

17. Let A and C be distinct points in the plane. Show that B is the midpoint of the segment $\overline{AC}$ if and only if

$$H_C\, H_B = H_B\, H_A.$$

Solution. Suppose that A and B are distinct points in the plane. We show first that the product $H_B\, H_A$ is a translation by the directed line segment $2\overline{AB}$.

In the figure below, let

$$P' = H_A(P) \qquad \text{and} \qquad P'' = H_B(P').$$

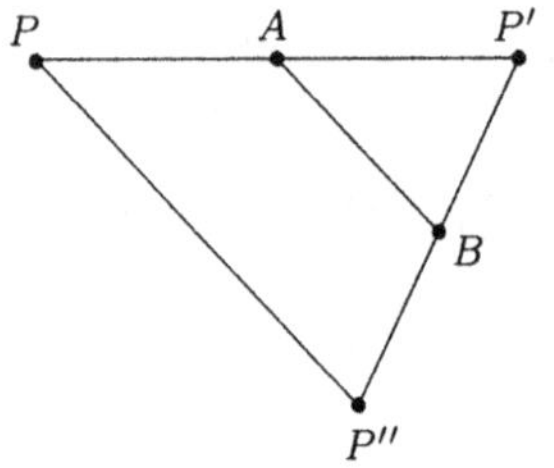

By the converse of the Midline Theorem, since A and B are the midpoints of sides PP' and $P'P''$, respectively, then PP'' is parallel to AB and twice as long as AB.

Thus, if A, B, and C are distinct points in the plane, then

$$H_C\, H_B = T_{2BC}$$

and

$$H_B\, H_A = T_{2AB}.$$

Now, B is the midpoint of the directed segment $\overline{AC}$ if and only if

$$\overline{AB} = \overline{BC},$$

that is, if and only if

$$H_C\, H_B = T_{2BC} = T_{2AB} = H_B\, H_A.$$

19. Using halfturns, prove that the diagonals of a parallelogram bisect each other.

    Hint. Show that if N is the midpoint of the diagonal AC of parallelogram $ABCD$ so that $H_A\, H_N = H_N\, H_C$, then N is the midpoint of BD also, that is, $H_D\, H_N = H_N\, H_B$.

    Solution. Since $ABCD$ is a parallelogram,

$$H_D\, H_C\, H_B\, H_A = I,$$

since

$$H_D\, H_C = T_{2CD}$$

and

$$\begin{aligned} H_B\,H_A &= T_{2AB} \\ &= T_{2DC} \\ &= (T_{2CD})^{-1} \\ &= (H_D\,H_C)^{-1}. \end{aligned}$$

Therefore,

$$H_B\,H_A = H_C\,H_D.$$

If N is the midpoint of the diagonal AC, then

$$H_A\,H_N = H_N\,H_C,$$

and so

$$H_B\,H_A\,H_N = H_B\,H_N\,H_C = H_C\,H_N\,H_B,$$

where the last equality follows from the fact that $H_B\,H_N\,H_C$ is an involution.

Therefore,

$$H_C\,H_D\,H_N = H_C\,H_N\,H_B$$

so that

$$H_D\,H_N = H_N\,H_B,$$

and N is also the midpoint of the diagonal BD.

21. Find all triangles such that three given noncollinear points are the midpoints of the sides of the triangle.

Hint. Given P, Q, and R, $H_R\,H_Q\,H_P$ fixes a vertex of a unique triangle $\triangle P'Q'R'$, as in the figure below.

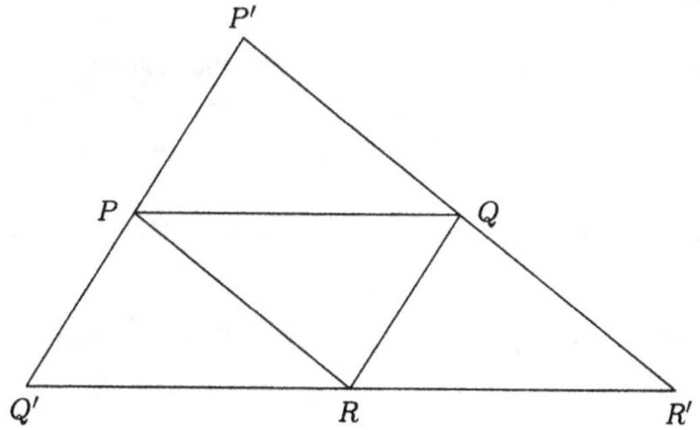

Solution. Suppose that $\triangle P'Q'R'$ is a triangle such that P, Q, and R are the respective midpoints of the sides $P'Q'$, $P'R'$, and $Q'R'$.

We have

$$H_Q\,H_R\,H_P(P') = H_Q\,H_R(Q') = H_Q(R') = P'$$

so that P' is the unique fixed point of the isometry (halfturn) $H_Q\, H_R\, H_P$.

Similarly,

$$H_R\, H_Q\, H_P(Q') = H_R\, H_Q(P') = H_R(R') = Q'$$

so that Q' is the unique fixed point of the isometry (halfturn) $H_R\, H_Q\, H_P$.

Finally,

$$H_R\, H_P\, H_Q(R') = H_R\, H_P(P') = H_R(Q') = R'$$

so that R' is the unique fixed point of the isometry (halfturn) $H_R\, H_P\, H_Q$.

Therefore, the triangle $\triangle P'Q'R'$ is uniquely determined by the three non-collinear points P, Q, and R.

23. Prove that if $R_n R_m$ fixes the point P and $m \neq n$, then the point P is on both lines m and n.

Solution. Suppose that P is a fixed point for the isometry $R_n R_m$, but P is not on both m and n. For example, suppose $P \notin m$.

Since $R_n R_m(P) = P$, then

$$R_n^2 R_m(P) = R_n(P),$$

that is,

$$R_m(P) = R_n(P).$$

Now, let

$$Q = R_m(P) = R_n(P),$$

and note that if $Q = P$, then $R_m(P) = P$. This implies that $P \in m$, which is a contradiction, and therefore $Q \neq P$.

Thus, m is the perpendicular bisector of the line segment joining $Q = R_m(P)$ and P, and n is also the perpendicular bisector of the line segment joining $Q = R_n(P)$ and P, which contradicts the fact that $m \neq n$.

Therefore, we must have $P \in m$. A similar argument shows that we must have $P \in n$ also.

25. Let m be a line with equation $2x + y = 1$. Find the equations of the transformation R_m.

Solution. If the equation of the line m is

$$ax + by + c = 0,$$

then the slope of m is $-a/b$, while the slope of a line $m_\perp$ that is perpendicular to m is b/a.

For a point P, let $P' = R_m(P)$ and suppose that P has Cartesian coordinates (x, y), while P' has Cartesian coordinates (x', y').

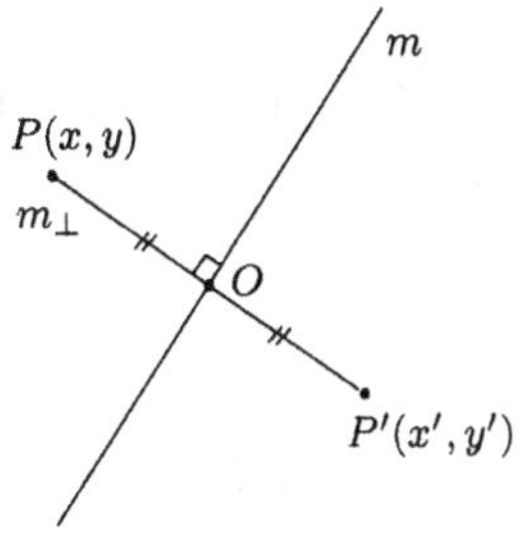

Since P and P' are on $m_\perp$, we have

$$y' - y = \frac{b}{a}(x' - x),$$

and since the midpoint O of PP' has coordinates $\left(\frac{x + x'}{2}, \frac{y + y'}{2}\right)$, and O is on the line m, then

$$a\left(\frac{x + x'}{2}\right) + b\left(\frac{y + y'}{2}\right) + c = 0.$$

Solving these equations for x' and y', the equations of the reflection R_m are given by

$$x' = x - \frac{2a}{a^2 + b^2}(ax + by + c)$$

$$y' = y - \frac{2b}{a^2 + b^2}(ax + by + c).$$

For the line $2x + y - 1 = 0$, we have $a = 2$, $b = 1$, and $c = -1$, so the equations of the reflection in this line are

$$x' = x - \frac{4}{5}(2x + y - 1)$$

$$y' = y - \frac{2}{5}(2x + y - 1).$$

27. Given triangles ABC and DEF, where $\triangle ABC \equiv \triangle DEF$ and where

$$A(0,0),\ B(5,0),\ C(0,10),\ D(4,2),\ E(1,-2),\ F(12,-4),$$

find the equations of the lines such that the product of reflections in the lines maps $\triangle ABC$ to $\triangle DEF$.

Solution. Note first that

$$AB = DE = 5, \qquad AC = DF = 10, \qquad \text{and} \qquad BC = EF = \sqrt{125}$$

so that $\triangle ABC \equiv \triangle DEF$ by the **SSS** congruency theorem.

Let ℓ be the perpendicular bisector of the segment AD. Since the midpoint of AD is the point

$$\frac{1}{2}(0+4, 0+2) = (2,1),$$

and AD has slope $-1/2$, the equation of ℓ is $y = -2x + 5$.

Therefore, the reflection R_ℓ has equations

$$x' = -\frac{3}{5}x - \frac{4}{5}y + 4$$
$$y' = -\frac{4}{5}x + \frac{3}{5}y + 2,$$

see Problem 8.25.

The images of the vertices of $\triangle ABC$ under the reflection R_ℓ are

$$A' = R_\ell(A) = (4,2) = D, \qquad B' = R_\ell(B) = (1,-2) = E$$

and

$$C' = R_\ell(C) = (-4,8),$$

as shown in the figure on the following page, and A' and B' are in the correct positions.

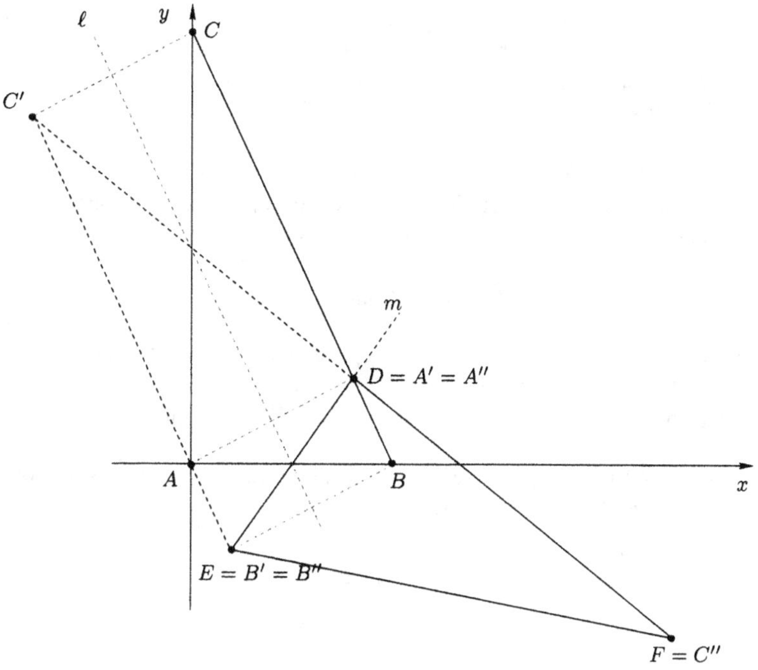

Now, let m be the perpendicular bisector of the segment CC''. Since the slope of m is the slope of DE, which is $4/3$, then the equation of m is $4x - 3y - 10 = 0$.

Therefore, the reflection R_m has equations

$$x'' = -\frac{7}{25}x' + \frac{24}{25}y' + \frac{16}{5}$$

$$y'' = \frac{24}{25}x' - \frac{7}{25}y' - \frac{12}{5},$$

again, see Problem 8.25. The images of the vertices of $\triangle A'B'C'$ under the reflection R_m are

$$A'' = R_m(A') = (4, 2) = D, \qquad B'' = R_m(B') = (1, -2) = E$$

and

$$C'' = R_m(C') = (12, -4) = F.$$

Therefore, the image of $\triangle ABC$ under the isometry

$$\alpha = R_m R_\ell$$

is the triangle $\triangle DEF$.

29. Let $A_0 = B_0$ be a given point and $\ell_1, \ell_2, \ldots, \ell_n$ be given lines. For $1 \le k \le n$, let A_k be obtained from A_{k-1} by a reflection across ℓ_k, and let B_k be obtained from B_{k-1} by a reflection across ℓ_{n-k+1}. For what values of n will A_n coincide with B_n?

Solution. When n is odd, A_n is obtained from A_0 by a reflection and B_n is obtained from B_0 by a reflection. Now, the composition of these two reflections is the composition of the $2n$ reflections across

$$\ell_1,\ \ell_2,\ \ldots,\ \ell_n,\ \ell_n,\ \ell_{n-1},\ \ldots,\ \ell_1.$$

Since they cancel in pairs, the composition is clearly the identity. This means that the composition of the two halfturns by which A_n and B_n are obtained is also the identity. It follows that A_n must coincide with B_n.

This is not necessarily the case when n is even, as illustrated by the counterexample in the diagram below, where $n = 2$.

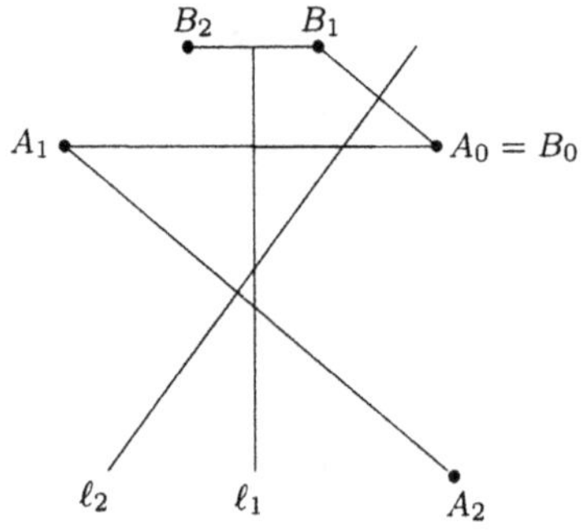

CHAPTER 9

THE PRODUCT OF DIRECT ISOMETRIES

1. If ℓ, m, and n are the perpendicular bisectors of the sides AB, BC, and CA, respectively, of $\triangle ABC$, then

$$T = R_n\, R_m\, R_\ell$$

is a reflection in which line?

Solution. Note that

$$R_\ell(A) = B, \qquad R_m(B) = C, \qquad \text{and} \qquad R_n(C) = A,$$

so that

$$R_n R_m R_\ell(A) = R_n R_m(B) = R_n(C) = A.$$

Solutions Manual to Accompany Classical Geometry: Euclidean, Transformational, Inversive, and Projective, First Edition. By I. E. Leonard, J. E. Lewis, A. C. F. Liu, G. W. Tokarsky.

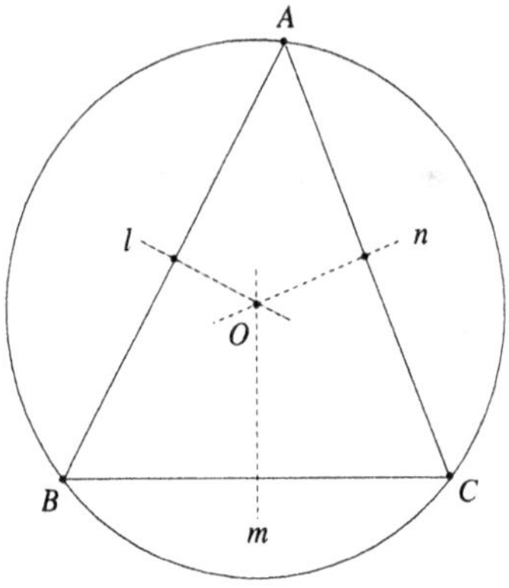

Also, since the lines ℓ, m, and n are concurrent at the circumcenter O, then

$$R_n R_m R_\ell(O) = R_n R_m(O) = R_n(O) = O.$$

Finally, since

$$\begin{aligned}(R_n R_m R_\ell)^2 &= R_n R_m R_\ell R_n R_m R_\ell = R_n R_m R_\ell R_\ell R_m R_n \\ &= R_n R_m R_m R_n = R_n R_n = I.\end{aligned}$$

Therefore, the isometry $T = R_n R_m R_\ell$ is an involution and has two fixed points A and O, thus is a reflection about the line through A and O.

3. Find Cartesian equations for lines m and n such that

$$R_m R_n(x, y) = (x + 2, y - 4).$$

Solution. Note that

$$R_m R_n(x, y) = (x + 2, y - 4) = T_{AB}(x, y)$$

is a translation by the directed line segment $\overline{AB}$ where

$$A = (0, 0) \qquad \text{and} \qquad B = (2, -4),$$

so that m and n are parallel, and both are perpendicular to $\overline{AB}$.

We can take the direction of the lines m and n to be $\overline{AC}$ where $C = (2, 1)$, and let m be the line with this direction passing through the point $(0, 0)$, and n be the line passing through the point $(2, -4)$, as in the figure on the following page,

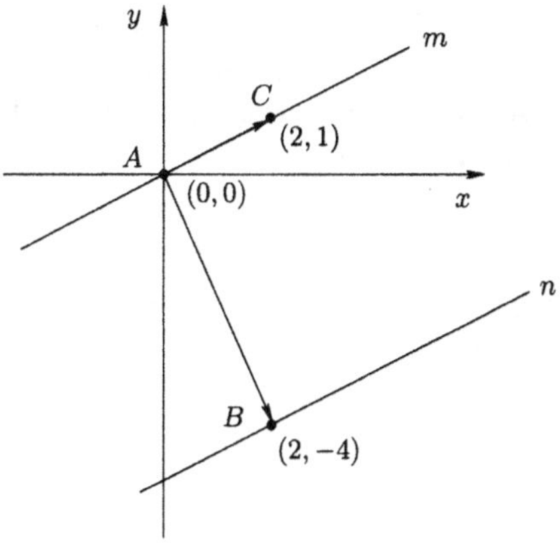

then

$$x = 2t$$
$$y = t \qquad (m)$$

and

$$x = 2 + 2t$$
$$y = -4 + t. \qquad (n)$$

5. Let C be a point on the line ℓ, and show that

$$R_\ell\, R_{C,\theta}\, R_\ell = R_{C,-\theta}.$$

Solution. Let A be an arbitrary point in the plane, and let B be a second point on the line ℓ. Let

$$A' = R_\ell(A), \qquad A'' = R_{C,\theta}(A'), \qquad \text{and} \qquad A''' = R_\ell(A''),$$

as shown in the figure.

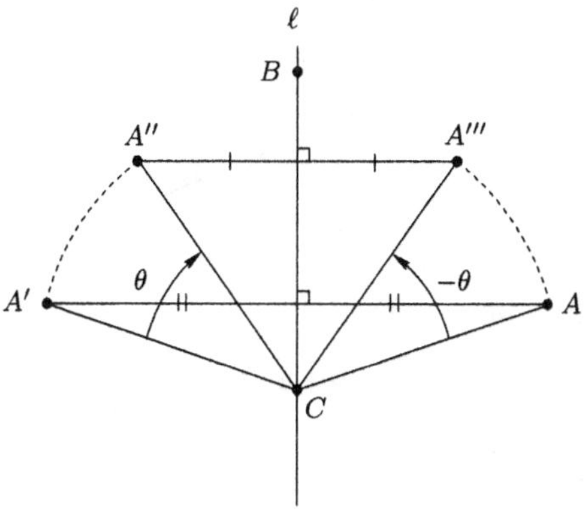

From the SAS congruence theorem, we have $CA = CA'$, and since $R_{C,\theta}$ is a rotation, we have $CA' = CA''$, and again by the SAS congruence theroem we have $CA'' = CA'''$, that is, the points A, A', A'', and A''' all lie on a circle with center at C and radius CA.

We have

$$\angle ACB - \angle A'''CB = |\,\theta\,| = \angle A'CB - \angle A''CB,$$

and the directed angles are

$$\angle ACA''' = -\theta = -\angle A'CA''.$$

Therefore

$$A''' = R_\ell R_{C,\theta} R_\ell(A) = R_{C,-\theta}(A),$$

and since A is arbitrary, we have

$$R_\ell R_{C,\theta} R_\ell = R_{C,-\theta}.$$

7. Show that if S, T, $T\,S$, and $T^{-1}\,S$ are rotations, then the centers of S, $T\,S$, and $T^{-1}\,S$ are collinear.

Solution. Let A and B be the centers of rotation for $S = R_{A,\theta}$ and $T = R_{B,\phi}$. We construct the center of rotation C for the product $T\,S$, that is,

$$R_{C,\theta+\phi} = R_{B,\phi}\,R_{A,\theta}$$

as follows.

Let ℓ be the line through the points A and B, and let m be the line through A making a directed angle of $\theta/2$ with the line ℓ, as shown in the figure.

Next draw the line n through B making a directed angle $\phi/2$ with the line ℓ, as in the figure. Since $T\,S$ is a rotation, then m and n are not parallel, so they intersect at a point C as shown.

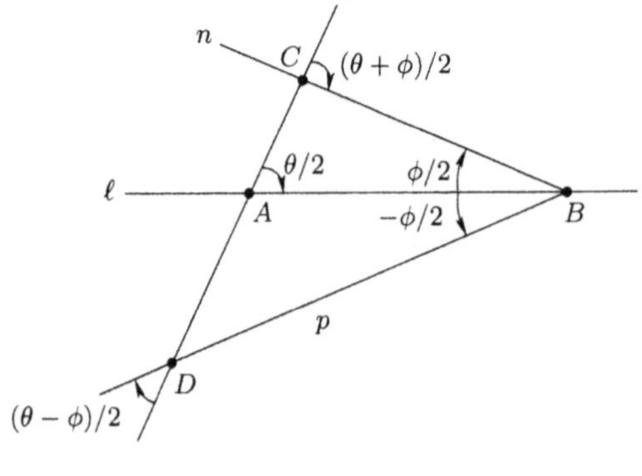

We construct the center of rotation D for the product $T^{-1}\,S$, that is,

$$R_{D,\theta-\phi} = R_{B,-\phi}\,R_{A,\theta}$$

in the same way.

Draw the line p through B making a directed angle $-\phi/2$ with the line ℓ, since $T^{-1}\,S$ is a rotation, then m and p are not parallel, so they intersect at a point D as shown.

The centers of rotation of S, $T\,S$, and $T^{-1}\,S$, that is, the points A, C, and D, are collinear and all lie on the line m.

9. Prove *Thomsen's Relation*: for any lines a, b, and c, we have

$$\begin{aligned} &R_c\,R_a\,R_b\,R_c\,R_a\,R_b\,R_a\,R_b\,R_c\,R_a\,R_b\,R_c \\ &\qquad\times R_b\,R_a\,R_c\,R_b\,R_a\,R_b\,R_a\,R_c\,R_b\,R_a = I, \end{aligned}$$

where I is the identity transformation.

Solution. First we note that $(R_cR_aR_b)^2$ is a translation.

If $a,\ b,\ c$ are concurrent or parallel, then $R_cR_aR_b = R_m$ is a reflection in a line m, and

$$(R_cR_aR_b)^2 = R_m^2 = I,$$

a translation by a directed line segment of length 0.

If $a,\ b,\ c$ are neither concurrent nor parallel, then $R_cR_aR_b$ is a glide reflection, say

$$R_cR_aR_b = T_{AB}R_\ell = R_\ell T_{AB},$$

where $\overline{AB}$ is parallel to ℓ, so that

$$(R_cR_aR_b)^2 = T_{AB}R_\ell R_\ell T_{AB} = T_{AB}^2R_\ell^2 = T_{AB}^2 = T_{2AB}.$$

We have

$$\begin{aligned} &R_c\,R_a\,R_b\,R_c\,R_a\,R_b\,R_a\,R_b\,R_c\,R_a\,R_b\,R_c\,R_b\,R_a\,R_c\,R_b\,R_a\,R_b\,R_a\,R_c\,R_b\,R_a \\ &\quad = R_c\,R_a\,R_b\,R_c\,R_a\,R_b\,R_a\,R_b\,R_c\,R_a\,R_b\,R_c \\ &\qquad\qquad \times R_b\,R_a\,R_c\,R_b\,R_a\,R_c\,R_c\,R_b\,R_a\,R_c\,R_b\,R_a \\ &\quad = (R_c\,R_a\,R_b)^2(R_a\,R_b\,R_c)^2(R_b\,R_a\,R_c)^2(R_c\,R_b\,R_a)^2. \end{aligned}$$

Since $(R_c\,R_b\,R_a)^2$, $(R_c\,R_a\,R_b)^2$, $(R_b\,R_a\,R_c)^2$, and $(R_a\,R_b\,R_c)^2$ are all translations, they commute.

Now note that

$$\begin{aligned}(R_c\,R_b\,R_a)^2(R_a\,R_b\,R_c)^2 &= R_c\,R_b\,R_a\,R_c\,R_b\,R_a\,R_a\,R_b\,R_c\,R_a\,R_b\,R_c\\ &= R_c\,R_b\,R_a\,R_c\,R_b\,R_b\,R_c\,R_a\,R_b\,R_c\\ &= R_c\,R_b\,R_a\,R_c\,R_c\,R_a\,R_b\,R_c\\ &= R_c\,R_b\,R_a\,R_a\,R_b\,R_c\\ &= R_c\,R_b\,R_b\,R_c\\ &= R_c\,R_c\\ &= I.\end{aligned}$$

Therefore,

$$(R_c\,R_b\,R_a)^2 = \left((R_a\,R_b\,R_c)^2\right)^{-1}$$

and

$$(R_c\,R_a\,R_b)^2 = \left((R_b\,R_a\,R_c)^2\right)^{-1},$$

so that

$$(R_c\,R_a\,R_b)^2(R_a\,R_b\,R_c)^2(R_b\,R_a\,R_c)^2(R_c\,R_b\,R_a)^2 = I,$$

and Thomsen's relation holds.

11. If $x' = ax + by + c$ and $y' = bx - ay + d$ with $a^2 + b^2 = 1$ are the equations for an isometry T, show that T is a reflection if and only if

$$ac + bd + c = 0 \qquad \text{and} \qquad ad - bc - d = 0.$$

Solution. First we show that if m is a line through the origin making a directed angle θ with the positive x-axis, and R_x is a reflection in the x-axis, then

$$R_{O,2\theta} = R_m\,R_x,$$

therefore $R_m = R_{O,2\theta}\,R_x$.

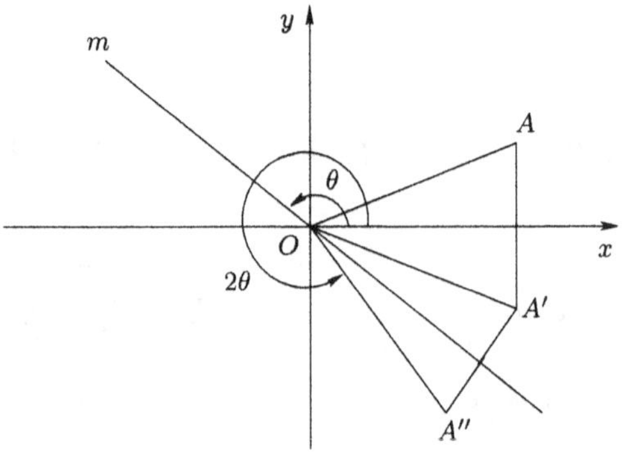

Let A be an arbitrary point in the plane and let $A' = R_x(A)$ and $A'' = R_m(A')$, so that $A'' = R_m R_x(A)$. From the figure we see that $A''(A) = R_{O,2\theta}(A)$, and since A is arbitrary then

$$R_{O,2\theta} = R_m\, R_x.$$

Since

$$R_x(x, y) = (x, -y)$$

and

$$R_{O,2\theta}(x, y) = (x\cos 2\theta - y\sin 2\theta, x\sin 2\theta + y\cos 2\theta),$$

then the equations of the reflection R_m are given by

$$\begin{aligned} x' &= x\cos 2\theta + y\sin 2\theta \\ y' &= x\sin 2\theta - y\cos 2\theta \end{aligned}$$

By translating, rotating, and translating back, the equations of a reflection in a line ℓ passing through the point (h, k) and making a directed angle θ with the positive x-axis are given by

$$\begin{aligned} x' &= (x - h)\cos 2\theta + (y - k)\sin 2\theta + h \\ y' &= (x - h)\sin 2\theta - (y - k)\cos 2\theta + k \end{aligned}$$

Now, if T is an isometry with equations $x' = ax + by + c$ and $y' = bx - ay + d$ with $a^2 + b^2 = 1$, by letting $a = \cos 2\theta$ and $b = \sin 2\theta$, then from the above we see that T is a reflection if and only if

$$c = h - ah - kb \qquad \text{and} \qquad d = k - bh + ka,$$

and this is the case if and only if

$$ac + bd + c = 0 \qquad \text{and} \qquad ad - bc - d = 0.$$

13. If the equations for the rotation $R_{P,\theta}$ are

$$2x' = -\sqrt{3}x - y + 2 \qquad \text{and} \qquad 2y' = x - \sqrt{3}y - 1,$$

find the center of rotation P and and the angle of rotation θ.

Solution. Writing the equations in the form

$$\begin{aligned} x' &= -\frac{\sqrt{3}}{2}x - \frac{1}{2}y + 1 \\ y' &= \frac{1}{2}x - \frac{\sqrt{3}}{2}y - \frac{1}{2}, \end{aligned}$$

we see that these are the equations of a rotation $R_{P,\theta}$ about the point $P = (h, k)$ through the angle θ, where

$$\cos\theta = -\frac{\sqrt{3}}{2} \qquad \text{and} \qquad \sin\theta = \frac{1}{2},$$

so that $\theta = \dfrac{5\pi}{6}$.

To find the point $P = (h, k)$ we note that since P is a fixed point of the rotation $R_{P,\theta}$, then

$$h = -\frac{\sqrt{3}}{2}h - \frac{1}{2}k + 1$$

$$k = \frac{1}{2}h - \frac{\sqrt{3}}{2}k - \frac{1}{2}$$

with solution

$$h = 1 - \frac{\sqrt{3}}{4} \qquad \text{and} \qquad k = \frac{3}{4} - \frac{\sqrt{3}}{2}.$$

15. If the isometry H_P is a halfturn, show that given any two perpendicular lines m and n that intersect at the point P, we have $H_P = R_m\, R_n$.

Solution. Given a halfturn H_P about the point P, if m and n are perpendicular lines that intersect at P, then $R_m R_n$ is a rotation about P by an angle of 180, that is, $R_m R_n = H_P$.

17. Given a line b and a point A, show that the following conditions are equivalent:

    (a) $A \in b$,
    (b) $H_A\, R_b = R_b\, H_A$,
    (c) $R_b(A) = A$,
    (d) $H_A(b) = b$,
    (e) $R_b\, H_A$ (or $H_A\, R_b$) is an involution,
    (f) $R_b\, H_A$ is a reflection in the line through A perpendicular to b.

Solution.

(a) implies (b). If $A \in b$, then

$$R_b H_A R_b = H_{R_b(A)} = H_A$$

since $R_b(A) = A$ if $A \in b$. Therefore,

$$H_A R_b = R_b H_A.$$

(b) implies (c). Suppose that $H_A R_b = R_b H_A$, then

$$R_b H_A R_b = H_A$$

and

$$R_b H_A R_b = H_{R_b(A)}$$

imply that $R_b(A) = A$.

(c) implies (d). Suppose that $R_b(A) = A$, since the only fixed points of R_b are on b, then $A \in b$. Now let P be an arbitrary point on b, then $H_A(P)$ is on the line joining A and P, so that $H_A(P) \in b$ also, therefore $H_A(b) \subseteq b$. However, $H_A(b)$ is a line, and so $H_A(b) = b$.

(d) implies (e). Suppose that $H_A(b) = b$, then

$$H_A R_b H_A = R_{H_A(b)} = R_b,$$

and so

$$H_A R_b H_A R_b = R_b H_A R_b H_A = R_b^2 = I,$$

that is, $H_A R_b$ and $R_b H_A$ are involutions.

(e) implies (f). Suppose that $H_A R_b$ is an involution, and let a be the line through A perpendicular to b.

Since $H_A R_b$ is an involution, then

$$H_A R_b H_A R_b(A) = I(A) = A,$$

so that

$$R_b H_A R_b(A) = H_A(A) = A,$$

and the halfturn $R_b H_A R_b = H_{R_b(A)}$ has A as a fixed point, and therefore $A = R_b(A)$, which implies that $A \in b$.

Now let $P \in a$ be arbitrary and let $P' = R_b(P)$, since b is perpendicular to a, then $P' \in a$ and $H_A(P') = P$, so that $H_A R_b(P) = P$ for all points $P \in a$, and $H_A R_b$ is a reflection in a. The inverse $R_b H_A$ is also a reflection in the line a.

(f) implies (a). Suppose that $R_b H_A$ is a reflection in the line a through A perpendicular to b. Let $b' = H_A(b)$, then the line b' is parallel to b, and since $A \in a$, then

$$A = R_b H_A(A) = R_b(A),$$

that is, A is a fixed point of R_b, and so $A \in b$.

19. Show that nonidentity rotations of the plane with different centers do not commute.

Solution. If C and D are points in the plane, and $\alpha = R_{D,\phi}$, then

$$\alpha R_{C,\theta} \alpha^{-1} = R_{\alpha(C),\theta},$$

so that

$$R_{D,\phi}\, R_{C,\theta} = R_{\alpha(C),\theta}\, R_{D,\phi}.$$

Now, if

$$R_{D,\phi}\, R_{C,\theta} = R_{C,\theta}\, R_{D,\phi},$$

then

$$R_{\alpha(C),\theta} = R_{C,\theta}.$$

This implies that C is the unique fixed point of $R_{\alpha(C),\theta}$, which implies that $\alpha(C) = C$, and so

$$R_{D,\phi}(C) = C,$$

which implies $D = C$.

Therefore, if $C \neq D$, then $R_{C,\theta}$ and $R_{D,\phi}$ do not commute.

21. Given a point A and two lines ℓ and m, construct a square $ABCD$ such that B lies on ℓ and D lies on m.

Solution. Given a line ℓ and a point A not on ℓ, we construct a square $ABCX$ with B on ℓ.

Draw the line p through A perpendicular to ℓ, hitting ℓ at F, and draw the line q through X perpendicular to p, hitting p at T, as in the figure below.

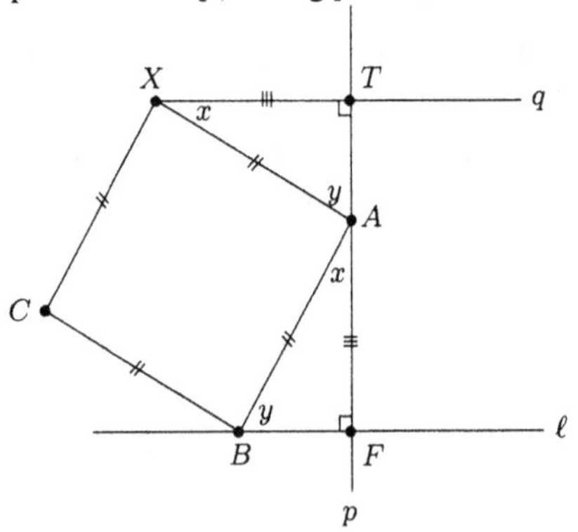

Since $AX = AB$, and $\angle AXT = \angle BAF$, then the right triangles $\triangle AFB$ and $\triangle XTA$ are congruent, and therefore $AF = XT$.

If A and ℓ are fixed, then as the vertex B of the square moves along the line ℓ, the distance of the vertex X from the line p is unchanged since $XT = AF$ is fixed, so that X moves along the line t which is parallel to and at a distance AF from p, as in the figure on the following page.

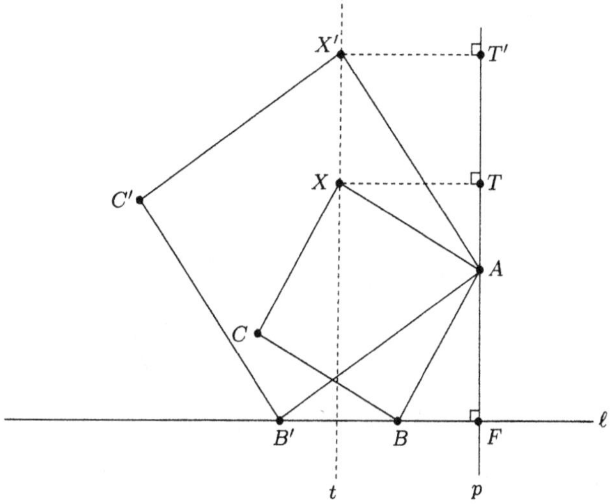

Now the construction is clear. As in the figure below, given lines ℓ and m and a point A not on ℓ or m, drop a perpendicular p from A to ℓ, hitting ℓ at F. Next draw a line t parallel to p and a distance AF from p, hitting m at D. Now that the side AD of the square $ABCD$ has been constructed, we erect a perpendicular from A hitting ℓ at B, and then erect perpendiculars from B and D intersecting at C.

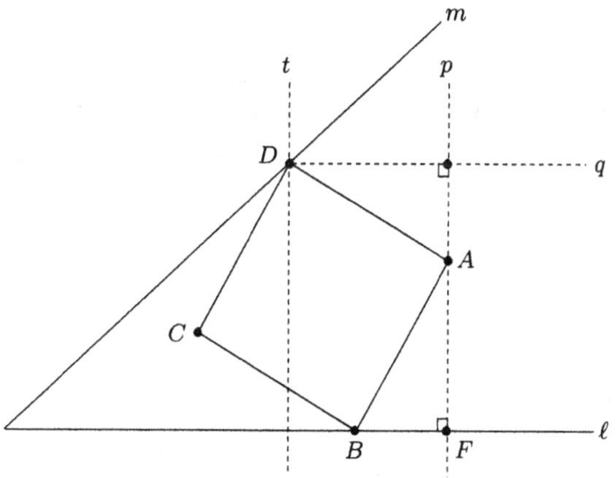

The case when A lies on either ℓ or m is left as an exercise.

23. Consider a triangle $\triangle ABC$ (oriented counterclockwise) with positive angles α, β, γ at A, B, C. Show that

$$R_{A,2\alpha}\, R_{B,2\beta}\, R_{C,2\gamma} = \mathbf{I}.$$

Are there other similar formulas?

Solution. In the figure,

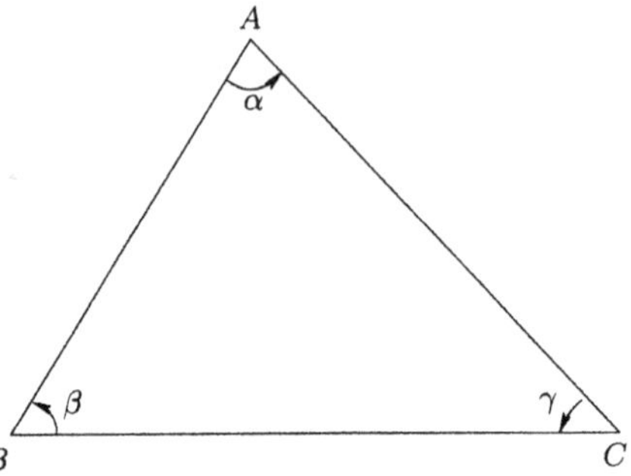

we have

$$R_{B,2\beta}\, R_{C,2\gamma} = R_{A,2(\beta+\gamma)},$$

so that

$$R_{A,2\alpha}\, R_{B,2\beta}\, R_{C,2\gamma} = R_{A,2\alpha}\, R_{A,2(\beta+\gamma)} = R_{A,2(\alpha+\beta+\gamma)} = R_{A,360} = I.$$

Similarly,

$$R_{C,2\gamma}\, R_{A,2\alpha}\, R_{B,2\beta} = I \qquad \text{and} \qquad R_{B,2\beta}\, R_{C,2\gamma}\, R_{A,2\alpha} = I.$$

25. Let perpendiculars erected at arbitrary points on the sides of triangle $\triangle ABC$ meet in pairs at points P, Q, and R. Show that the triangle PQR is similar to the given triangle.

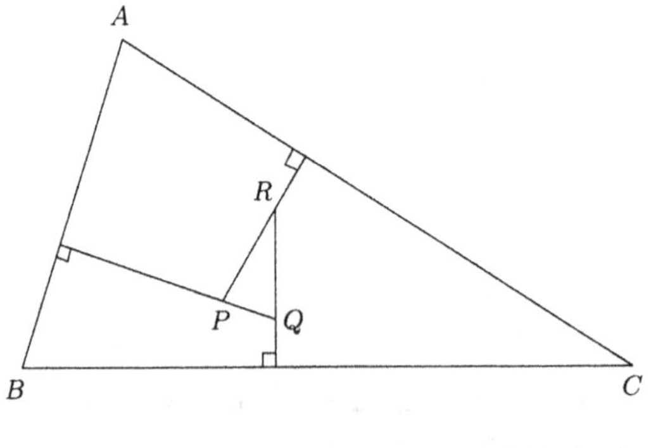

Solution. A 90° rotation of $\triangle ABC$ into $\triangle A'B'C'$ leaves the sides of $\triangle A'B'C'$ parallel to the corresponding sides of $\triangle PRQ$.

By the AAA similarity theorem,

$$\triangle A'B'C' \sim \triangle PQR,$$

and since the rotation is an isometry, then

$$\triangle ABC \equiv \triangle A'B'C',$$

so that $\triangle ABC \sim \triangle PQR$.

27. Given points A, B, and P in the plane, construct the reflection of P in the line AB using a Euclidean compass alone.

Solution. Draw the circle $\mathcal{C}(A, AP)$ and the circle $\mathcal{C}(B, BP)$, and let P' be their second intersection point. By the SSS congruency condition, we have $\triangle APB \equiv \triangle AP'B$, so that the quadrilateral $APBP'$ is a kite, and hence its diagonals are prependicular.

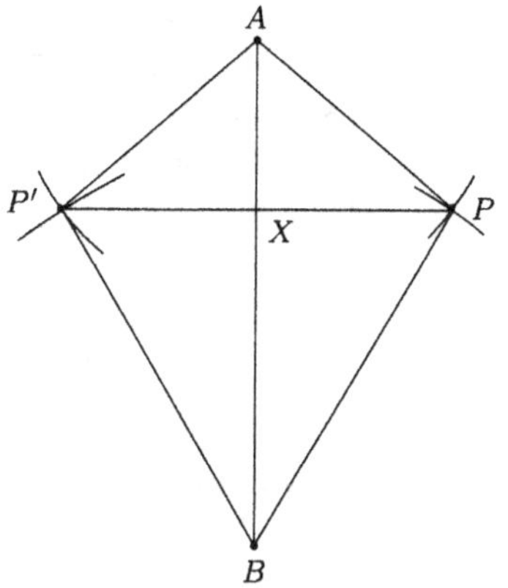

By the HSR congruence condition, $\triangle AXP \equiv AXP'$, so $PX = P'X$, and AB is the perpendicular bisector of PP'. Therefore,

$$P' = R_{AB}(P),$$

that is, P' is the reflection of the point P in the line AB.

29. Let $A_0B_0C_0$ be an equilateral triangle with the vertices in clockwise order. We first rotate it 60° counterclockwise about A_0 to obtain $A_1B_1C_1$, then about B_1 to obtain $A_2B_2C_2$, and finally about C_2 to obtain $A_3B_3C_3$. We continue to rotate about A_3, B_4, C_5, and so on, until $A_nB_nC_n$ occupies the same physical space as $A_0B_0C_0$. What is the minimum positive value of n?

Solution. The diagram below shows that the minimum value is $n = 6$. The center of the hexagon is $A_2 = A_5 = B_0 = B_3 = C_1 = C_4$.

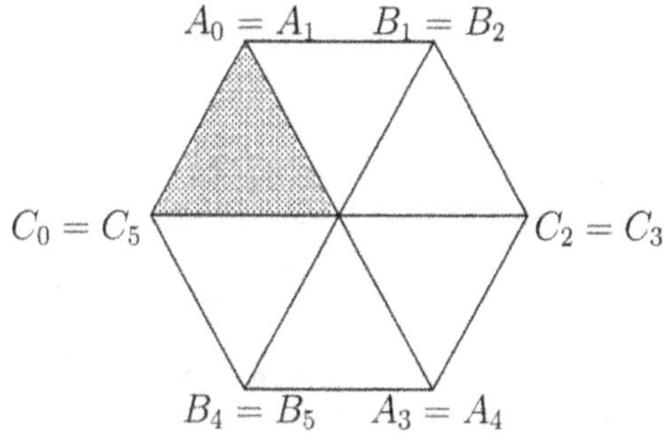

31. A halfturn about a point O is a $180°$ rotation about the point O. Prove that the composition of:

    (a) two halfturns is a translation or the identity,

    (b) a translation and a halfturn is a halfturn.

Solution.

(a) Obviously, if the two halfturns are about the same point, the composition will be the identity. Consider a segment A_0B_0 which is mapped into A_1B_1 by a halfturn about some point O_1. Since $A_0O_1 = O_1A_1$ and $B_0O_1 = O_1B_1$, $A_0B_0A_1B_1$ is a parallelogram. Suppose a halfturn about a point O_2 maps A_1B_1 into A_2B_2, and then $A_1B_1A_2B_2$ is also a parallelogram. It follows that $A_0B_0B_2A_2$ is a parallelogram, and A_2B_2 is obtained from A_0B_0 by a translation. See the diagram below on the left.

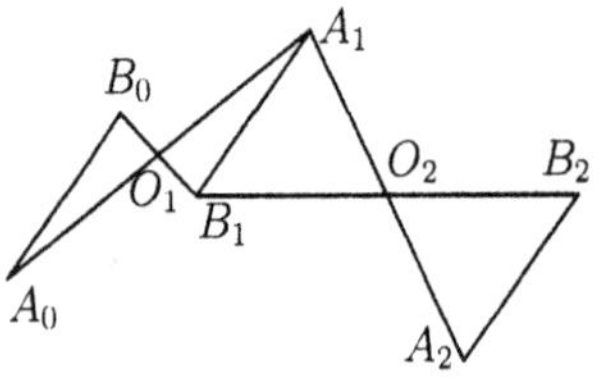

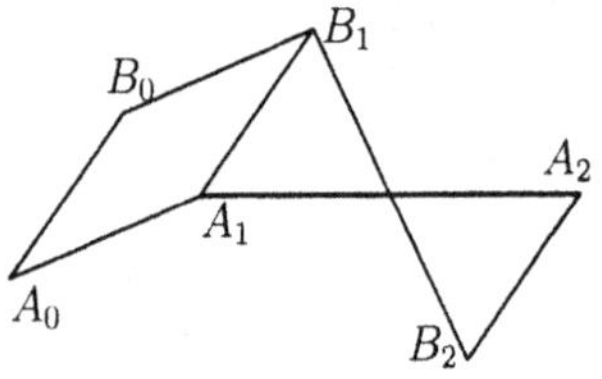

(b) Consider a segment A_0B_0 mapped into A_1B_1 by some translation, so that $A_0B_0B_1A_1$ is a parallelogram. Suppose some halfturn maps A_1B_1 into A_2B_2, then $A_1B_1A_2B_2$ is also a parallelogram. It follows that $A_0B_0A_2B_2$ is a parallelogram, and A_2B_2 is obtained from A_0B_0 by a halfturn about the center of this parallelogram. See the diagram above on the right.

33. Let A_0, B_0, O_1, O_2, $\ldots$, O_n be given points. For $1 \le k \le n$, let A_kB_k be obtained from $A_{k-1}B_{k-1}$ by a halfturn about O_k.

    (a) Prove that $A_0A_n = B_0B_n$ if n is even.

    (b) What conclusion may be drawn if n is odd?

Solution.

(a) Since the composition of an even number of halfturns is a translation, A_nB_n is the image of A_0B_0 under a translation. It follows that

$$A_0A_n = B_0B_n.$$

(b) Since the composition of an odd number of halfturns is a halfturn, A_nB_n is the image of A_0B_0 under a half-turn. We may conclude that

$$A_0B_n = B_0A_n.$$

35. Let $A_0 = B_0$, O_1, O_2, …, O_n be given points. For $1 \le k \le n$, let A_k be obtained from A_{k-1} by a halfturn about O_k, and let B_k be obtained from B_{k-1} by a halfturn about O_{n-k+1}. For what values of n will A_n coincide with B_n?

Solution. When n is odd, A_n is obtained from A_0 by a halfturn and B_n is obtained from B_0 by a halfturn. Now, the composition of these two halfturns is the composition of the $2n$ halfturns about

$$O_1,\ O_2,\ \ldots,\ O_n,\ O_n,\ O_{n-1},\ \ldots,\ O_1.$$

Since they cancel in pairs, the composition is clearly the identity. This means that the composition of the two halfturns by which A_n and B_n are obtained is also the identity. It follows that A_n must coincide with B_n. This is not necessarily the case when n is even, as illustrated by the counterexample in the diagram below where $n = 2$.

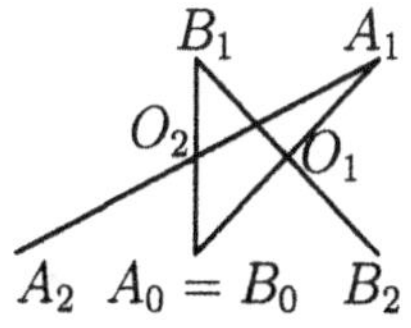

CHAPTER 10

SYMMETRY AND GROUPS

1. Prove that a finite group of isometries cannot contain two halfturns about distinct points.

   Solution. If the finite group of isometries $\mathcal{G}$ contains two halfturns H_P and H_Q, where $P \neq Q$, then $\mathcal{G}$ also contains the product

$$H_P H_Q = T_{2PQ}.$$

   However, this is impossible since a finite group of isometries contains only reflections and rotations.

Solutions Manual to Accompany Classical Geometry: Euclidean, Transformational, Inversive, and Projective, First Edition. By I. E. Leonard, J. E. Lewis, A. C. F. Liu, G. W. Tokarsky.

3. Prove that if a triangle is invariant under a reflection, then the triangle must be isosceles.

Solution. Suppose that $\triangle ABC$ is invariant under the reflection R_ℓ, then since vertices are mapped to vertices and not all vertices are on the line ℓ, we may assume that $R_\ell(B) = C$.

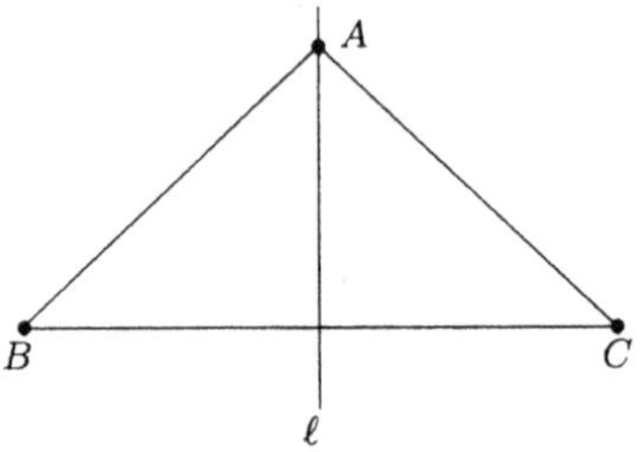

Thus, $\angle B = \angle C$, and by the converse of the Isosceles Triangle Theorem, $\triangle ABC$ is isosceles.

5. If $H_{O_1} H_{O_2} = H_{O_2} H_{O_1} = T$, prove that $T = I$, the identity transormation.

Solution. We note that

$$H_{O_1} H_{O_2} = T_{2O_2O_1} \quad \text{and} \quad H_{O_2} H_{O_1} = T_{2O_1O_2},$$

so that

$$\overline{O_2O_1} = \overline{O_1O_2} = -\overline{O_2O_1},$$

and $\overline{O_2O_1} = 0$, and hence $O_1 = O_2$.

Therefore,

$$H_{O_1} H_{O_2} = H_{O_2} H_{O_1} = T_{2O_2O_2} = I,$$

the identity transformation.

7. Find a plane figure P such that its group of symmetries equal

   (a) the dihedral group D_2 of order 4,

   (b) the dihedral group D_1 of order 2.

Solution.

(a) One such example is the figure on the right, with group of symmetries

$$\mathcal{D}_2 = \{\, I,\ R_\ell,\ R_{\ell'},\ H_O \,\}$$

where ℓ is the line through A and B, O is the midpoint of the segment AB, and $R_{\ell'}$ is the line through O perpendicular to ℓ.

To see that this is the case: if

$$T : \triangle ABC \to \triangle ABC,$$

then $T = I$, while if

$$T : \triangle ABC \to \triangle ABC',$$

then $T = R_\ell$, while if

$$T : \triangle ABC \to \triangle BAC,$$

then $T = R_{\ell'}$, while if

$$T : \triangle ABC \to \triangle BAC',$$

then $T = H_O$.

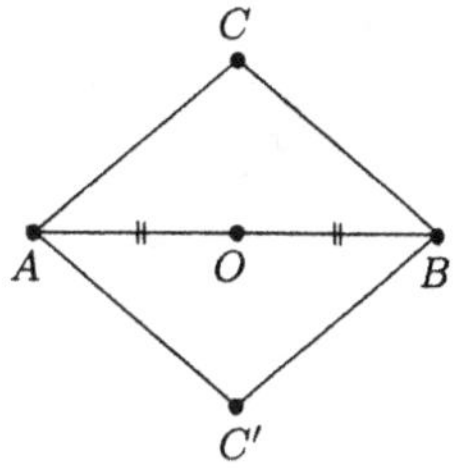

(b) One such example is the figure on the right, any nonequilateral isosceles triangle has symmetry group

$$\mathcal{D}_1 = \{ I, R_\ell \}$$

where ℓ is the line through A perpendicular to the side BC.

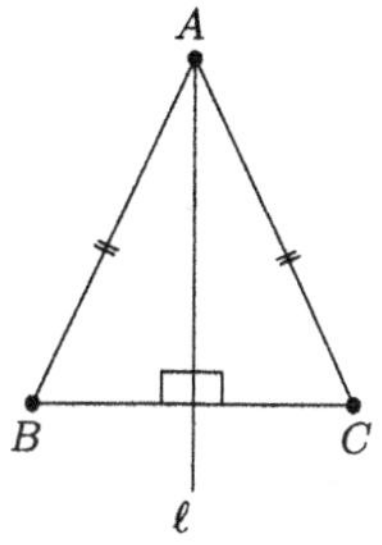

To see that this is the case: if a symmetry T satisfies

$$T(A) = A, \quad T(B) = B, \quad T(C) = C,$$

then $T = I$, the identity transformation, while if T satisfies

$$T(A) = A, \quad T(B) = C, \quad T(C) = B,$$

then $T = R_\ell$, that is, T is a reflection in the line ℓ.

9. Let $\mathcal{G}$ be a group of isometries whose subgroup of translations is generated by T_{AB}, where $AB \neq 0$. Prove that if $R_\ell \in \mathcal{G}$, then either $\overline{AB}$ is parallel to ℓ, or $\overline{AB}$ is perpendicular to ℓ.

Solution. We first note that

$$T_{AB}R_\ell = R_\ell T_{A'B'}$$

where $A'B'$ is the image of AB under R_ℓ. Let P be an arbitrary point in the plane, and let

$$R_\ell(P) = P'' \qquad \text{and} \qquad T_{AB}(P'') = P'$$

and

$$T_{A'B'}(P) = P''' \qquad \text{and} \qquad R_\ell(P''') = P''''.$$

As in the figure on the following page, ℓ is the perpendicular bisector of both segments $\overline{P'P'''}$ and $\overline{P'''P''''}$, so that $P'''' = P'$.

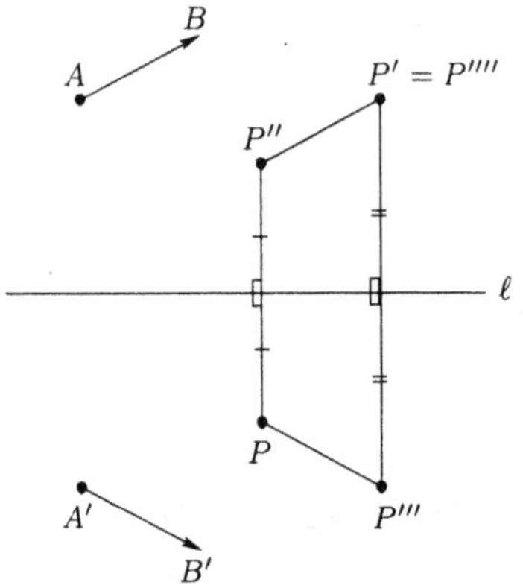

This means that

$$T_{A'B'} = R_\ell T_{AB} R_\ell \in \mathcal{G},$$

and hence

$$\overline{A'B'} = n\overline{AB}$$

for some integer n. However, $|A'Bp| = |AB|$ since it is a reflected image.

Hence, either

$$\overline{A'B'} = \overline{AB}, \quad \text{which can only happen if} \quad \overline{AB} \parallel \ell,$$

or if

$$\overline{A'B'} = -\overline{AB}, \quad \text{which can only happen if} \quad \overline{AB} \perp \ell.$$

Thus, either $\overline{AB}$ is parallel to ℓ, or $\overline{AB}$ is perpendicular to ℓ.

11. If a and b are elements of a group $\mathcal{G}$ and

$$(ab)^2 = a^2b^2,$$

show that $ab = ba$.

Solution. Let e be the identity element in the group $\mathcal{G}$, f a and b are elements of $\mathcal{G}$ such that $(ab)^2 = a^2b^2$, then

$$abab = (ab)^2 = a^2b^2 = aabb,$$

and multiplying this equation on the left by a^{-1} and on the right by b^{-1}, we obtain

$$(a^{-1}a)ba(bb^{-1}) = (a^{-1}a)ab(bb^{-1}),$$

so that

$$ebae = eabe.$$

Therefore, since $ea = ae = a$ and $eb = be = b$, then $ba = ab$.

13. A *cuboid* is a rectangular parallelopiped; that is, a parallelopiped where each plane face is orthogonal to four other faces and parallel to the fifth, as in the figure.

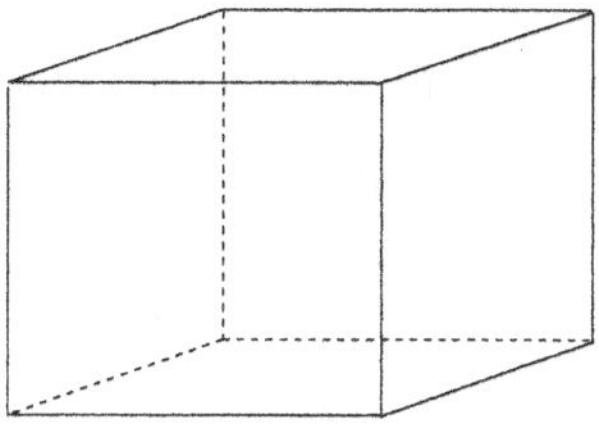

Find the group of symmetries of a cuboid with three unequal sides.

Solution. The symmetries of the cuboid can be written as:

- The identity transformation.
- Rotations by 180° degrees about three axes. The three rotations are about lines that are perpendicular to opposite faces and pass through the center of the faces, one of which is indicated in the figure below.

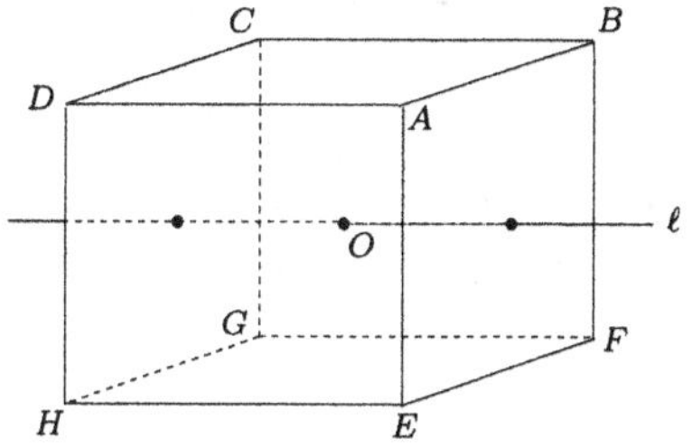

- Reflections in three mutually orthogonal planes passing through the center O parallel to the faces of the cuboid, one of which is indicated in the figure below.

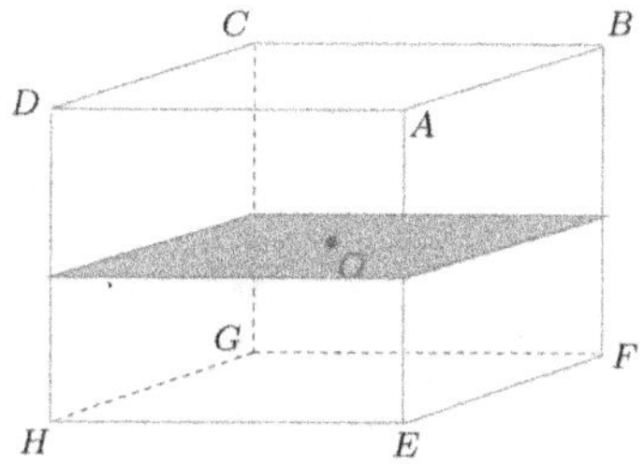

- A point reflection or inversion through the center O; that is, for each point P in the cuboid, the image is P', where O, P, and P' are collinear and $OP = OP'$.

15. Let $\mathcal{T}$ be a (nonequilateral) isosceles triangle.

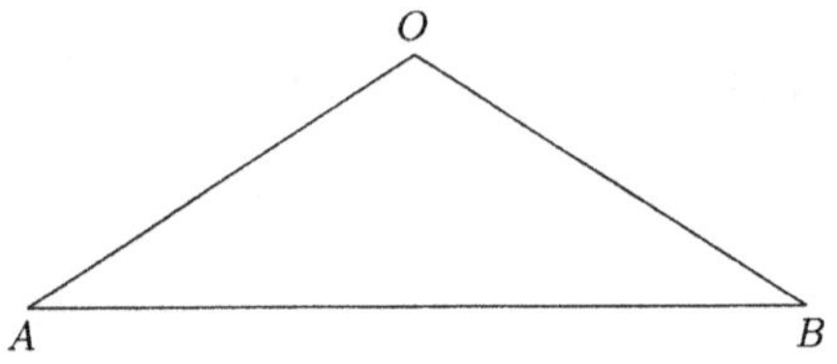

Find the group of symmetries of $\mathcal{T}$ and construct the Cayley table for the group.

Solution. Let m be the perpendicular bisector of the side AB, then the only symmetries of $\mathcal{T}$ are I, the identity, and R_m, a relfection in the line m. The Cayley table for the group $\mathcal{G} = \{ I, R_m \}$ is

$\cdot$	I	R_m
I	I	R_m
R_m	R_m	I

17. Prove that if $H_{O_0} \in \mathcal{F}$, a frieze group with translation subgroup

$$\mathcal{T} = \{T_{nAB}\} = \langle T_{AB} \rangle ,$$

then $H_{O_{n/2}} \in \mathcal{F}$ for every integer n, where $\overline{O_0 O_{n/2}} = \frac{n}{2}\overline{AB}$.

Solution. Since $H_{O_0} \in \mathcal{F}$ and $T_{nAB} \in \mathcal{T} \subset \mathcal{F}$, then $H_{O_0} T_{nAB} \in \mathcal{F}$. However, as can be seen from the figure on the following page, since

$$H_{O_0} = R_\ell R_{\ell'_0}$$

and

$$T_{nAB} = R_{\ell'_0} R_{\ell'_{n/2}},$$

then

$$\begin{aligned} H_{O_0} T_{nAB} &= R_\ell R_{\ell'_0} R_{\ell'_0} R_{\ell'_{n/2}} \\ &= R_\ell R_{\ell'_{n/2}} \\ &= H_{O_{n/2}}, \end{aligned}$$

and so $H_{O_{n/2}} \in \mathcal{F}$ for every integer n, where $\overline{O_0 O_{n/2}} = \frac{n}{2}\overline{AB}$.

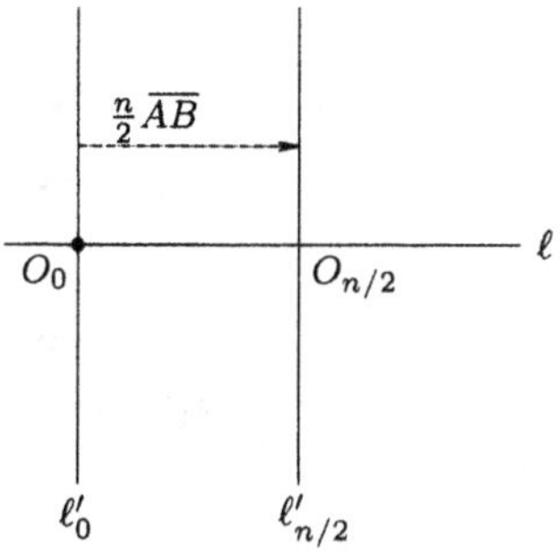

19. Find a different set of two generators for the frieze group

$$\mathcal{F} = \left\langle H_{O_{1/4}}, G_{\ell, \frac{1}{2}AB} \right\rangle,$$

where $\ell \parallel \overline{AB}$, and O_0 and $O_{\frac{1}{4}}$ are two fixed points on ℓ with $\overline{O_0 O_{\frac{1}{4}}} = \frac{1}{4}\overline{AB}$, as in the figure below.

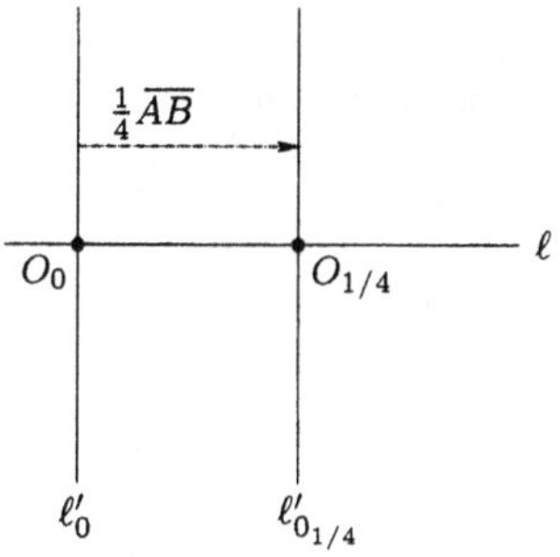

Solution. Let

$$\mathcal{F}_1 = \left\langle H_{O_{1/4}}, R_{\ell_0'} \right\rangle \qquad \text{and} \qquad \mathcal{F}_2 = \left\langle R_{\ell_0'}, G_{\ell, \frac{1}{2}AB} \right\rangle$$

where ℓ_0' and $\ell_{\frac{1}{4}0}'$ pass through O and $O_{\frac{1}{4}}$, respectively, and are perpendicular to ℓ. We will show that both $\mathcal{F}_1$ and $\mathcal{F}_2$ are equal to $\mathcal{F}$.

Note that

$$\begin{aligned} G_{\ell, \frac{1}{2}AB} R_{\ell_0'} &= R_\ell T_{AB} R_{\ell_0'} \\ &= R_\ell R_{\ell_{0_{1/4}}'} R_{\ell_0'} R_{\ell_0'} \\ &= R_\ell R_{\ell_{0_{1/4}}'} \\ &= H_{O_{1/4}}, \end{aligned}$$

that is,

$$H_{O_{1/4}} = G_{\ell, \frac{1}{2}AB} R_{\ell_0'},$$

and from this we have

$$G_{\ell,\frac{1}{2}AB} = H_{O_{1/4}} R_{\ell_0'}.$$

Therefore, both generators of $\mathcal{F}_1$ are in $\mathcal{F}_2$, and conversely, so that $\mathcal{F}_1 = \mathcal{F}_2$.

Finally, since

$$R_{\ell_0'} = H_{O_{1/4}} G_{\ell,\frac{1}{2}AB} \in \mathcal{F},$$

then both generators of $\mathcal{F}_1$ are in $\mathcal{F}$. Similarly, both generators of $\mathcal{F}$ are in $\mathcal{F}_1 = \mathcal{F}_2$, so that $\mathcal{F} = \mathcal{F}_1 = \mathcal{F}_2$.

21. Find the frieze group of an infinite horizontal strip consisting of repeated I's if the I's lie above the midline of the strip as shown below.

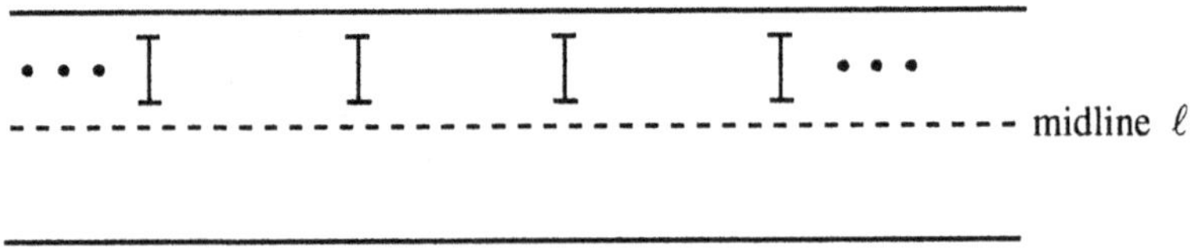

Solution. The frieze group of the pattern above is

$$\mathcal{F} = \langle T_{AB}, R_{\ell_0'} \rangle$$

where, as in the figure below O_0 and O_1 are points on ℓ with $\overline{O_0O_1} = \overline{AB}$.

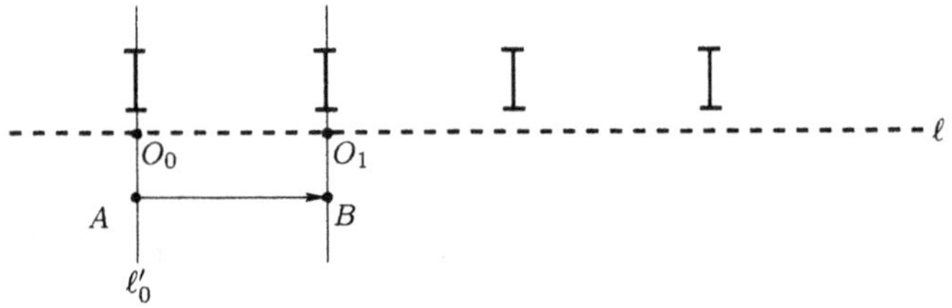

23. Consider a particular vertex of a square. Under a symmetry of the square, this vertex can land on any one of the four vertices. The remaining vertices must follow in order, either counterclockwise or clockwise. Thus, there are only $4 \times 2 = 8$ symmetries for the square. They are: the identity I, a 180° rotation or halfturn R, a 90° counterclockwise rotation A, a 90° clockwise rotation C, and four reflections H, V, D, and U, as shown in the figure below.

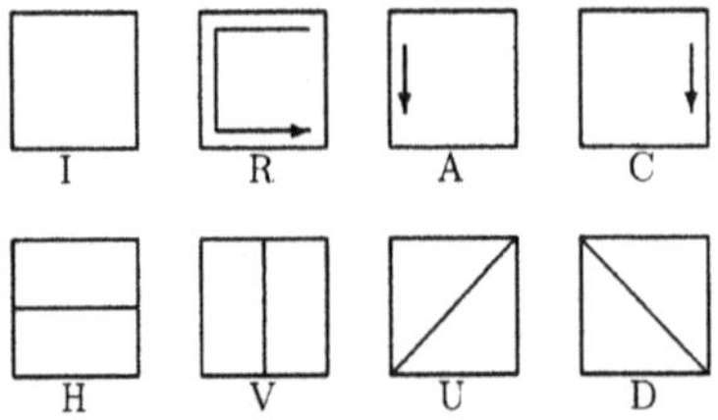

These eight symmetries of the square form a group under the operation of composition, the dihedral group of the square. Construct the operation table of this group.

Solution. The operation table is shown below.

	I	R	A	C	H	V	U	D
I	I	R	A	C	H	V	U	D
R	R	I	C	A	V	H	D	U
A	A	C	R	I	D	U	H	V
C	C	A	I	R	U	D	V	H
H	H	V	U	D	I	R	A	C
V	V	H	D	U	R	I	C	A
U	U	D	V	H	C	A	I	R
D	D	U	H	V	A	C	R	I

25. The complex numbers 1, -1, i, and $-i$ form a group under multiplication. Construct the operation table of this group.

Solution. The operation table is shown below.

$\times$	1	-1	i	$-i$
1	1	-1	i	$-i$
-1	-1	1	$-i$	i
i	i	$-i$	-1	1
$-i$	$-i$	i	1	-1

27. Let

$$a(x) = \frac{1}{1-x},$$
$$b(x) = \frac{x-1}{x},$$
$$c(x) = 1-x,$$
$$d(x) = \frac{x}{x-1},$$
$$e(x) = x,$$
$$f(x) = \frac{1}{x}.$$

These functions form a group with respect to the operation of composition. Construct the operation table.

Solution. The operation table is shown on the following page.

$\circ$	a	b	c	d	e	f
a	b	e	d	f	a	c
b	e	a	f	c	b	d
c	f	d	e	b	c	a
d	c	f	a	e	d	b
e	a	b	c	d	e	f
f	d	c	b	a	f	e

CHAPTER 11

HOMOTHETIES

1. Find the two centers of homothety for the top and bottom of an isosceles trapezoid.

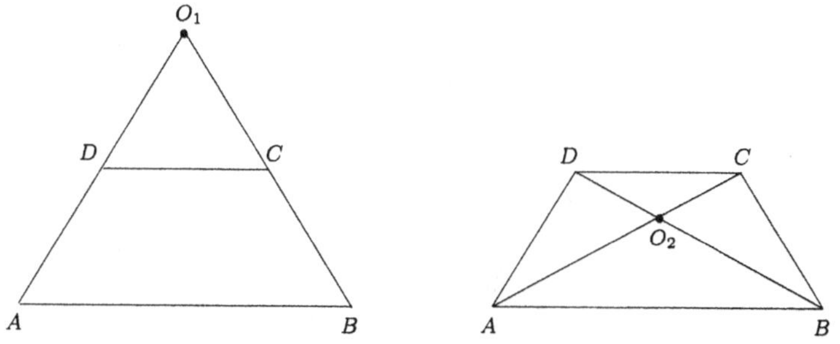

Solution. The centers of homothety are O_1 and O_2, as in the figure above. This follows immediately from the fact that parallel lines preserve proportions in triangles.

Solutions Manual to Accompany Classical Geometry: Euclidean, Transformational, Inversive, and Projective, First Edition. By I. E. Leonard, J. E. Lewis, A. C. F. Liu, G. W. Tokarsky.

3. Show that the product of two homotheties with the same center is a homothety and find its center and ratio.

Solution. Given two homotheties with the same center O, ratios k_1 and k_2, then for any point P in the plane,

$$H_{O,k_1}H_{O,k_2}(P) = H_{O,k_1}(P')$$

where $\overline{OP'} = k_2\overline{OP}$, and

$$H_{O,k_1}H_{O,k_2}(P) = H_{O,k_1}(P') = P''$$

where $\overline{OP''} = k_1\overline{OP'} = k_1k_2\overline{OP}$.

Note that O, P, P', and P'' are collinear, and therefore,

$$H_{O,k_1}H_{O,k_2}(P) = H_{O,k_1k_2}(P)$$

for all points P in the plane, that is, $H_{O,k_1}H_{O,k_2} = H_{O,k_1k_2}$. Thus, the product of two homotheties with the same center is a homothety with the same center and ratio the product of the two ratios.

5. Show that if $\triangle ABC$ and $\triangle A'B'C'$ are similar, with AB parallel to $A'B'$, AC parallel to $A'C'$, and BC parallel to $B'C'$, then the lines joining corresponding vertices are concurrent, and there is a homothety $\mathbf{H}(O,k)$ such that $\triangle A'B'C'$ is the image of $\triangle ABC$ under $\mathbf{H}(O,k)$. Find the center O and the ratio k.

Solution. Note first that if the triangles are congruent, then we can take the homothety to have center A and ratio 1, and this is a translation by $\overline{AA'}$.

Suppose that triangles $\triangle ABC$ and $\triangle A'B'C'$ are noncongruent similar triangles as in the figure below.

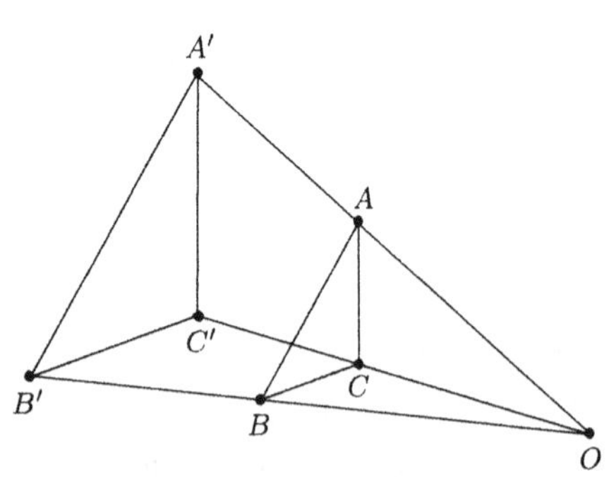

Since pairs of corresponding sides are parallel, then the two lines from each pair meet at a point at infinity; that is, $\triangle ABC$ and $\triangle A'B'C'$ are coaxial. By Desargues' Two Triangle Theorem, the triangles are also copolar at a point O.

Thus, lines joining corresponding vertices are concurrent at O, and the point O is the center of homothety with ratio

$$k = \frac{\overline{OA'}}{\overline{OA}} = \pm\frac{A'B'}{AB}.$$

Therefore, $\triangle A'B'C'$ is the image of $\triangle ABC$ under the homothety $H_{O,k}$.

7. Show that the inverse of the homothety $H(O,k)$ is the homothety $H(O,1/k)$.

Solution. Given the homothety $H(O,k)$ where $k > 0$, let P be an arbitrary point in the plane, then

$$H_{O,k}(P) = P'$$

where $\overline{OP'} = k\overline{OP}$.

Also,

$$H_{O,1/k}(P') = P''$$

where $\overline{OP''} = \frac{1}{k}\overline{OP'}$.

Hence,

$$\overline{OP''} = \frac{1}{k}\overline{OP'} = \frac{1}{k}k\overline{OP} = \overline{OP},$$

and by the property of directed distances, since O, P, P', and P'' are all collinear, this implies that $P'' = P$.

Therefore,

$$H_{O,1/k}H_{O,k}(P) = P$$

for all points P in the plane, so that $H_{O,1/k}H_{O,k} = I$, the identity transformation, and $H_{O,1/k} = H_{O,k}^{-1}$.

9. Show that a product of three homotheties is a homothety or a translation.

Solution. Let $\mathcal{G}$ be the set consisting of all homotheties and all translations. Since $\mathcal{G}$ is a set of transformations it contains the identity, $H_{O,1} = I$. The associative law holds, and for each $T \in \mathcal{G}$ the inverse $T^{-1} \in \mathcal{G}$ also. Thus, in order to show that $\mathcal{G}$ is a group we only have to show that it is closed under multiplication.

In the solution to Problem 11-4 we showed that the product of two homotheties H_{O_1,k_1} and H_{O_2,k_2} is either a homothety or a translation depending on whether the product of the ratios k_1k_2 is different from 1 or equal to 1, respectively. Theorem 9.3.1 showed that the product of two translations is either a translation or the identity. To finsh the proof that $\mathcal{G}$ is a group, we have to show that the product of a homothety and a translation is a homothety.

Given the homothety $H_{O,k}$ and the translation T_{AB}, let Q be a fixed point in the plane and let P be an arbitrary point. We consider the effect of the product $T_{AB}H_{O,k}$ on the point P. Let

$$P'' = H_{O,k}(P) \qquad \text{and} \qquad Q'' = H_{O,k}(Q)$$

and

$$P' = T_{AB}(P'') \qquad \text{and} \qquad Q' = T_{AB}(Q''),$$

as in the figure below.

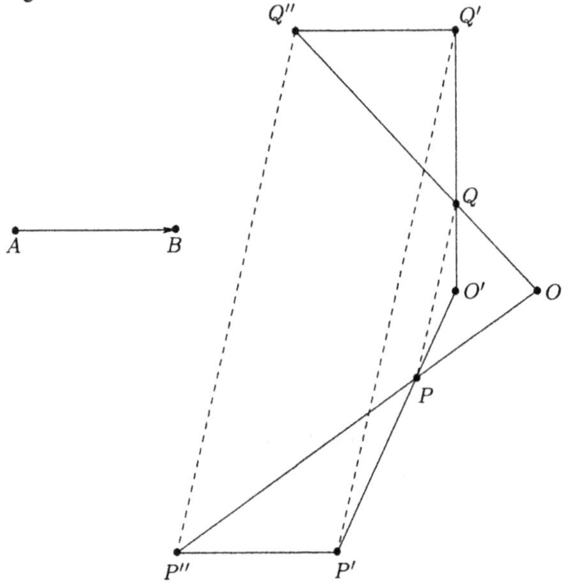

From the definition of a homothety, the points O, Q, Q'' are collinear, with

$$\overline{OQ''} = k\overline{OQ}.$$

Similarly, the points O, P, and P'' are collinear, with

$$\overline{OP''} = k\overline{OP}.$$

Now, $P'P''$ and $Q'Q''$ are parallel (they are both parallel to AB) and

$$P'P'' = AB = Q'Q'',$$

and since

$$\frac{OQ''}{OQ} = k = \frac{OP''}{OP},$$

then by the **sAs** similarity theorem, $\triangle OPQ \sim \triangle OP''Q''$ so that $PQ \parallel P''Q''$, and they are both parallel to $P'Q'$ since $T_{AB}(P''Q'') = P'Q'$.

Now, parallel lines perserve proportions in triangles, so that if O' is the point of intersection of PP' and QQ', then

$$\frac{O'P'}{O'P} = \frac{O'Q'}{O'Q} = k',$$

that is,

$$H_{O',k'}(P) = P' = T_{AB}H_{O,k}(P)$$

for all points P in the plane. Therefore, we have $T_{AB}H_{O,k} = H_{O',k'}$. In a similar way the product $H_{O,k}T_{AB}$ can be seen to be a homothety.

Now that we know that $\mathcal{G}$ is a group, it follows that the product of three homotheties is either a homothety or a translation.

11. $\overline{A_1B_1}$ and $\overline{A_2B_2}$ are two nonparallel segments such that $A_1B_1 = 2A_2B_2$, as in the figure below.

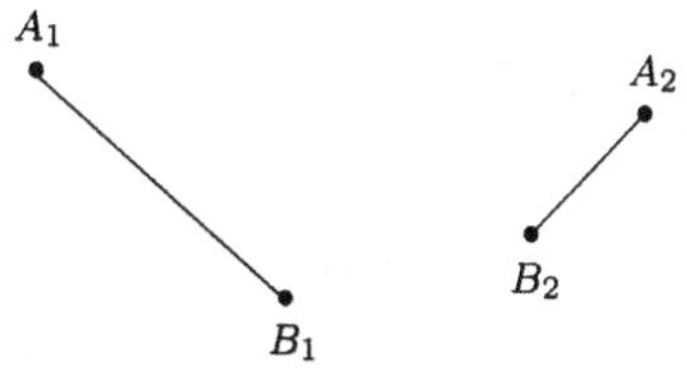

(a) Find a point O such that $\overline{A_2B_2}$ may be obtained from $\overline{A_1B_1}$ by means of a homothety centered at O with ratio $\frac{1}{2}$, followed by a rotation about O.

(b) Find a line ℓ and a point O on ℓ such that $\overline{A_2B_2}$ may be obtained from $\overline{A_1B_1}$ by a homothety centered at O with ratio $\frac{1}{2}$, followed by a reflection across ℓ.

Solution.

(a) Let P be the point of intersection of A_1B_1 and A_2B_2. Construct the circumcircles of triangles PA_1A_2 and PB_1B_2. If these two circles are tangent to each other at P, we may take O to be P and perform a $0°$ rotation after the homothety.

Otherwise, let O be the other point of intersection of the two circumcircles, as in the figure on the following page.

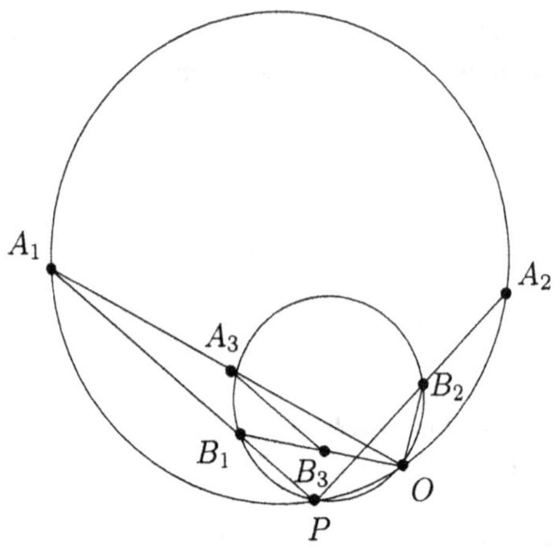

We then have

$$\angle A_1OA_2 = \angle A_1PA_2 = \angle B_1PB_2 = \angle B_1OB_2.$$

Denote the common measure of these angles by θ, then a rotation about O through an angle θ will take A_2B_2 to A_3B_3 with A_3 and B_3 on OA_1 and OB_1, respectively.

Moreover,

$$\angle OA_1B_1 = \angle OA_1P = \angle OA_2P = \angle OA_2B_2.$$

Hence triangles OA_1B_1 and OA_2B_2 are similar, so that $OA_1 = 2OA_3$ and $OB_1 = 2OB_3$.

Thus, a homothety centered at O with ratio $\frac{1}{2}$ will take A_1B_1 to A_3B_3, and a rotation about O through an angle $-\theta$ will take A_3B_3 to A_2B_2.

(b) Let P be the point of intersection of A_1B_1 and A_2B_2. Let Q be the point on B_1B_2 such that $QB_1 = 2QB_2$, then ℓ is the line through P parallel to the bisector of $\angle B_1PB_2$, as in the figure below.

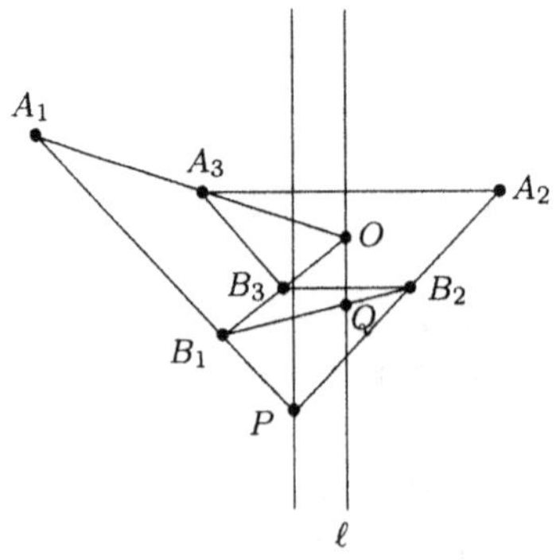

It follows that a reflection across ℓ will take A_2B_2 to A_3B_3 with A_3B_3 parallel to A_1B_1.

Let O be the point of intersection of ℓ with B_1B_3. Since the distance from B_3 to ℓ is equal to the distance from B_2 to ℓ, which is in turn half the distance from B_1 to ℓ since $QB_1 = 2QB_2$, then B_3 is the midpoint of OB_1.

Hence A_3 is the midpoint of OA_3; thus, a homothety centered at O with ratio $\frac{1}{2}$ will take A_1B_1 to A_3B_3, and a reflection across ℓ will take A_3B_3 to A_2B_2.

13. Construct a triangle ABC given its circumcenter O, its orthocenter H, and the vertex A.

Solution. Construct the centroid G of $\triangle ABC$ which lies on its Euler line OH with $HG = GO$, as in the figure below.

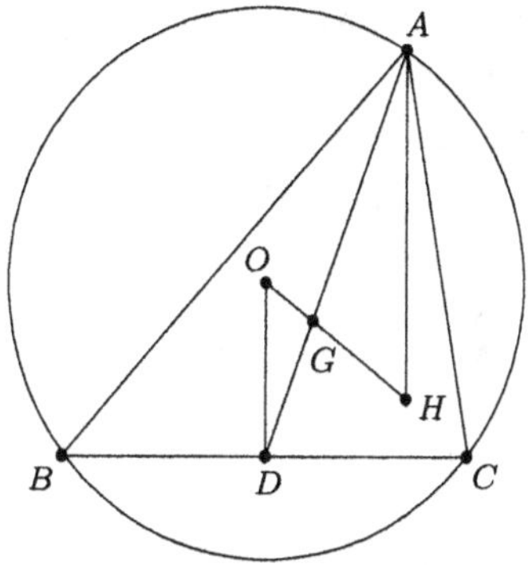

Extend AG to D so that $AG = 2GD$, and draw a line through D perpendicular to OD, cutting the circle with center O and radius OA at B and C.

Note that a homothety $H_{G,-1/2}$ centered at G with ratio $-\frac{1}{2}$ maps H onto O and maps A onto D. Since AH is perpendicular to BC, then OD is also perpendicular to BC, and $\triangle ABC$ is indeed the desired triangle.

15. Let P be a point on side BC of an equilateral triangle ABC, closer to C than B. Construct a point Q on CA and a point R on AB such that $\angle RPQ = 90°$ and $PR = 2PQ$.

Solution. In the figure on the following page, perform a $90°$ rotation about P, mapping AC onto A_1C_1, then perform the homothety $H_{P,2}$ centered at P with ratio 2, mapping A_1C_1 to A_2C_2.

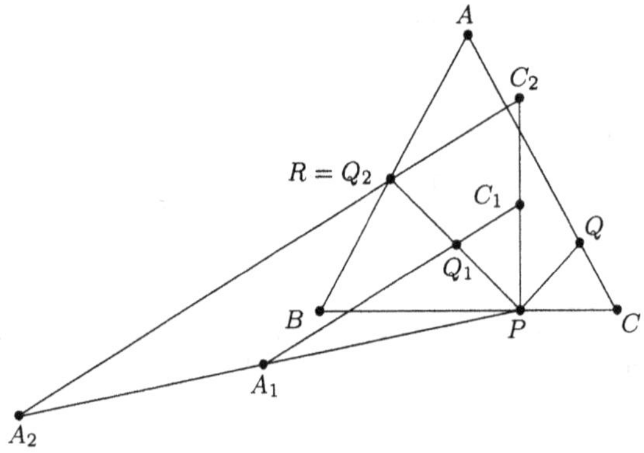

The point $R = Q_2$ is the point of intersection of A_2C_2 with AB. The point Q_1 is the midpoint of the segment PQ_2, and Q is obtained from Q_1 by a $-90°$ rotation about P, as shown in the figure above.

Note that if A_2C_2 and AB do not intersect, then there are no solutions.

17. The circles ω_1 and ω_2 intersect at M and N. The point A_1 is a variable point on ω_1, the point A_2 is the point of intersection of the line A_1M with ω_2, and the point B is the third vertex of an equilateral triangle A_1A_2B, with the vertices in the counterclockwise order. Prove that the locus of B is a circle.

Solution. There are two cases to consider, depending on whether M is between A_1 and A_2 or not, as in the figure below.

We claim that the shape of $\triangle A_1NA_2$ is independent of the choice of A_1A_2.

From Thales' Theorem, the chord MN subtends a constant angle α_1 on the major arc MN of ω_1, and a constant angle α_2 on the major arc of MN on ω_2.

- If M lies between A_1 and A_2, as in the figure above on the left, then

$$\angle A_1NA_2 = 180° - \angle NA_1A_2 - \angle NA_2A_1 = 180° - \alpha_1 - \alpha_2.$$

- If M does not lie between A_1 and A_2, as in the figure above on the right, then

$$\angle A_1NA_2 = \angle MA_2N - \angle MA_1N = 180° - \alpha_2 - \alpha_1.$$

Thus, the claim is justified.

Since N is a fixed point inside the equilateral triangle A_1A_2B, then $\angle A_2NB$ has a fixed measure θ. Now a rotation about N through an angle θ followed by a homothety centered at N with the constant ratio NB/NA_2 will map A_2 onto B. Since the locus of A_2 is ω_2, then the locus of B is also a circle.

19. Two circles ω_1 and ω_2 are tangent to each other at the point T. A line through T intersects ω_1 at A_1 and intersects ω_2 at A_2. Prove that the tangent to ω_1 at A_1 is parallel to the tangent to ω_2 at A_2.

Solution. Let the centers of ω_1 and ω_2 be O_1 and O_2, respectively. We may obtain ω_2 from ω_1 by a homothety centered at T with ratio

$$k = \begin{cases} -\dfrac{O_2T}{O_1T}, & \text{if the tangency is external} \\ \dfrac{O_2T}{O_1T}, & \text{if the tangency is internal.} \end{cases}$$

as in the figure below.

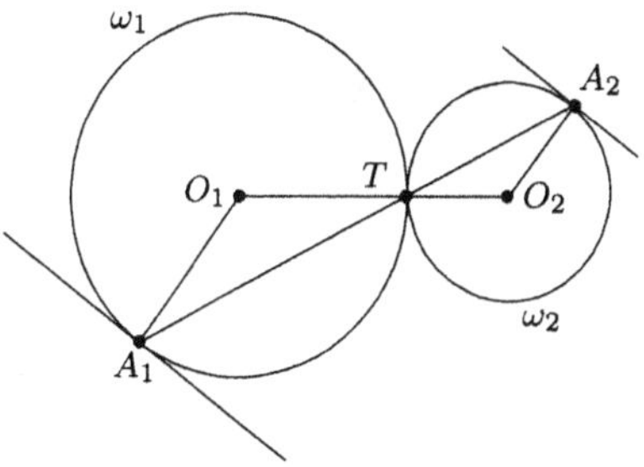

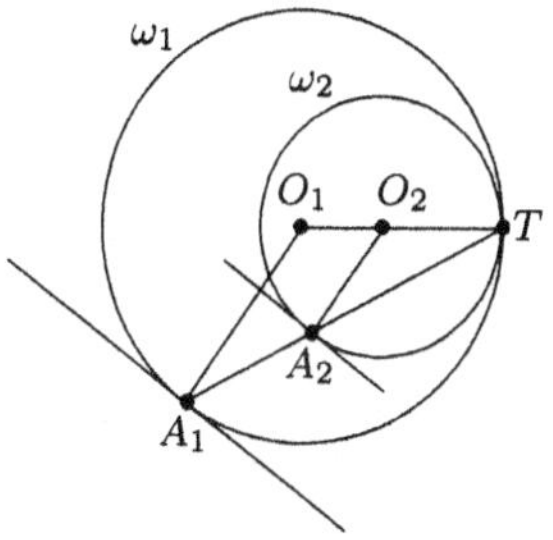

In either case, this homothety maps A_1 onto A_2 and maps the tangent to ω_1 at A_1 onto the tangent to ω_2 at A_2. Thus, the two tangents are parallel to each other.

21. The figure $ABCD$ is a quadrilateral, with AB parallel to DC. The extensions of DA and CB intersect at the point P, and the diagonals intersect at the point Q. Prove that PQ passes through the midpoints M and N of AB and CD, respectively.

Solution. As in the figure on the following page, the homothety centered at P with ratio CD/AB will map CD onto AB, and the midpoint N of CD onto the midpoint M of AB. Hence, P lies on the line MN.

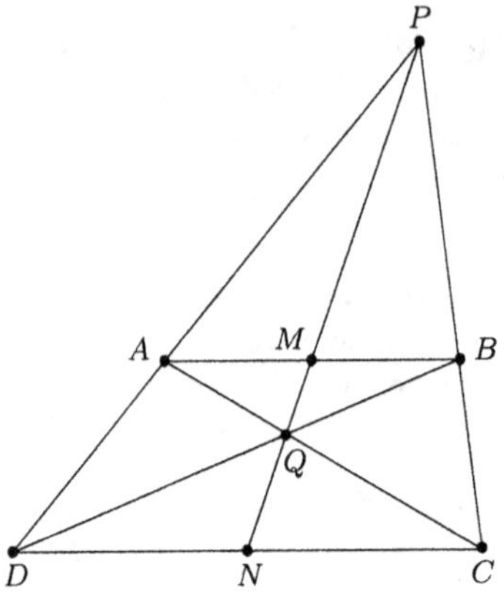

A homothety centered at Q with ratio $-CD/AB$ will map CD onto AB and N onto M. Hence, Q also lies on the line MN. Therefore, PQ passes through the midpoints M and N of AB and CD.

CHAPTER 12

TESSELLATIONS

1. There are twelve ways in which five unit squares can be joined edge to edge. The resulting figures are called ***pentominoes***, as shown in the figure on the following page.

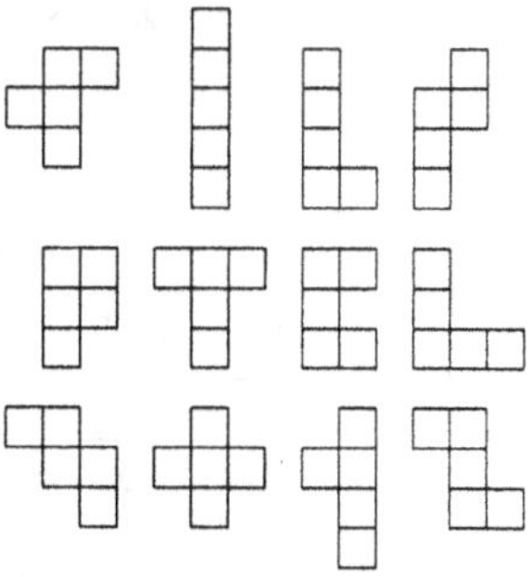

They are given letter names, F, I, L, N, P, T, U, V, W, X, Y and Z, respectively. Identify those that can tile a rectangle.

Solutions Manual to Accompany Classical Geometry: Euclidean, Transformational, Inversive, and Projective, First Edition. By I. E. Leonard, J. E. Lewis, A. C. F. Liu, G. W. Tokarsky.

Solution. The pentominoes that can tile a rectangle are the I-pentomino, the L-pentomino, the P-pentomino and the Y-pentomino, as shown in the figure below.

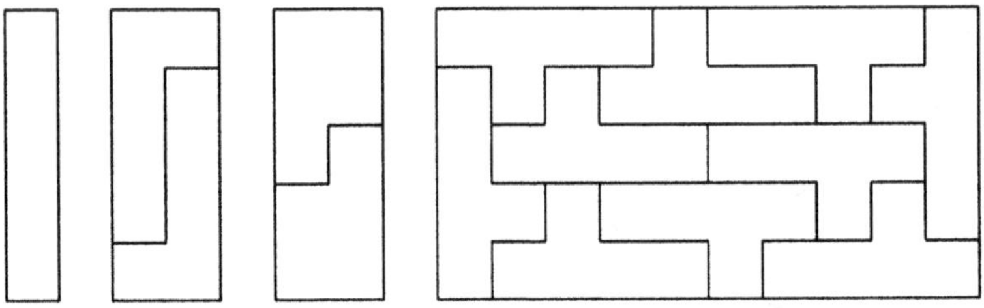

3. Prove that a pentomino that can tile a bent strip can also tile some infinite strip.

Solution. Divide one arm of the tiled bent strip into blocks of length 10. Within each block, there exists a path from the top edge of the arm to the bottom edge of the arm without cutting any of the pentominoes. Now, the number of such paths must be finite. Suppose there are n such paths. By taking $n + 1$ blocks, we must have two paths that are identical. The part of the tessellation between these two paths may be used to tile an infinite strip.

5. Obviously, a pentomino that can tile an infiinite strip can also tile the plane. Identify those that can tile the plane but cannot tile an infinite strip.

Solution. The pentominoes that can tile the plane but cannot tile and infinite strip are the T-pentomino, the U-pentomino, the X-pentomino and the Z-pentomino. The figure below shows that none can tile the bottom edge of an infinite strip. Squares marked A are inaccessible. While squares marked B can be covered, they lead right away to other inaccessible squares.

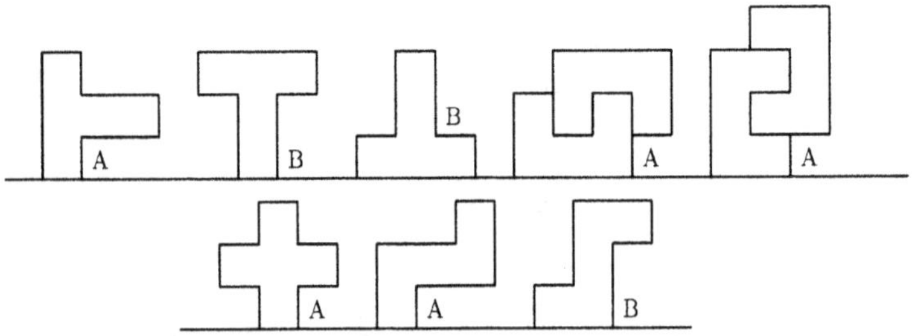

The figure on the following page shows that each can tile the plane.

7. (a) Show that the figure below can tile the plane.

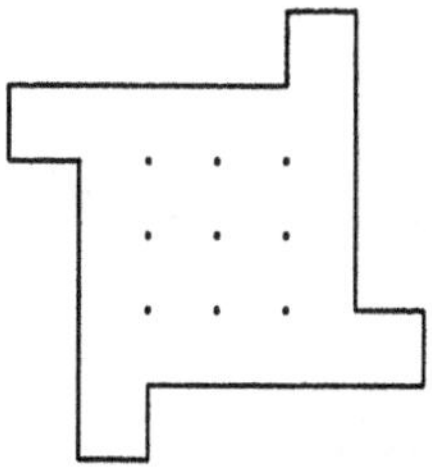

(b) For each of the P-pentomino, Y-pentomino, and Z-pentomino, dissect it into three pieces and reassemble them into a square.

Solution.

(a) The figure on the following page shows the desired tessellation.

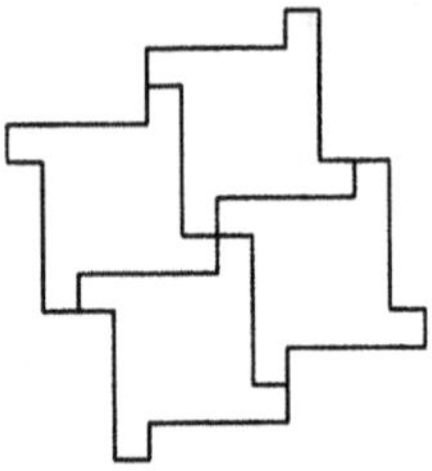

(b) Each of the three pentominoes can tile the figure in (a), and hence can tile the plane. When this tessellation is superimposed on the square tessellation, the solution to the dissection puzzle emerges.

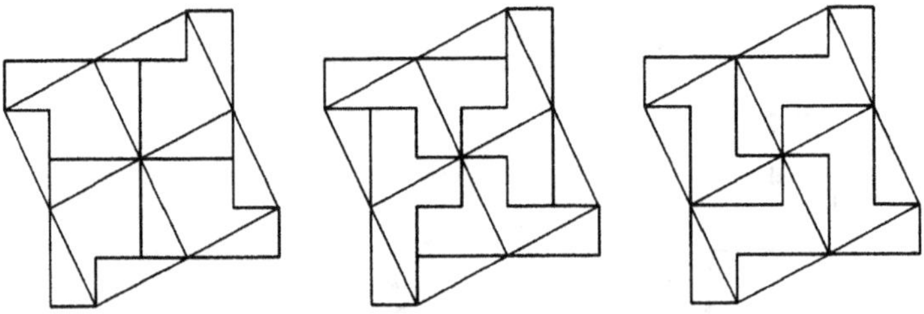

9. Dissect the figure below into three pieces and reassemble them into a square.

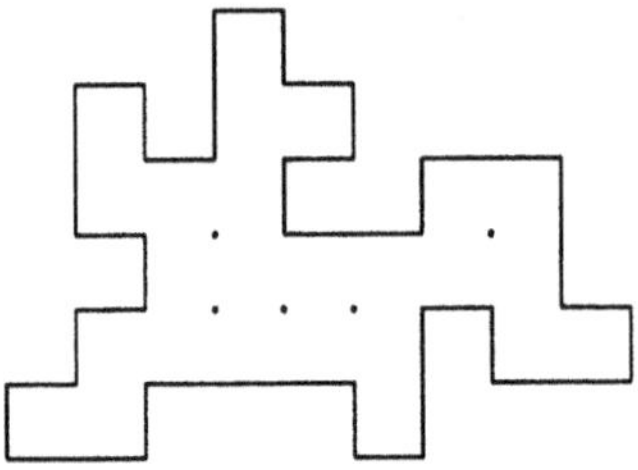

Solution. The figure below shows that figure given above can tile the plane. The solution to the dissection puzzle emerges when we superimpose this tessellation with the square tessellation.

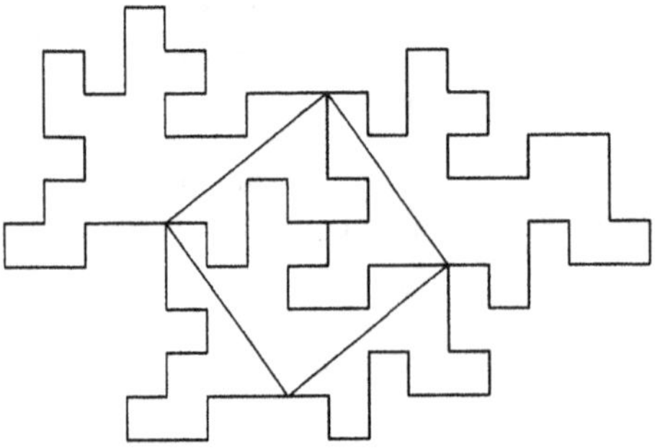

11. (a) Prove that the sum of measures of the exterior angles of a regular n-gon is 360° for any $n \geq 3$.

    (b) Use part (a) to prove that the sum of the measures of the central angles of a regular n-gon is given by $(n-2)180°$.

Solution.

(a) Consider a vector moving along one side of the regular polygon. Let the head of the vector pass through the far vertex of the side and wait until the tail of the vector reaches it. Next, pivot around this vertex so that the vector is now in the direction of the adjacent side, it has turned through an exterior angle of the polygon. Continue in this manner until the vector returns to its original position, so that it has made a 360° turn. In the process, it has turned through each exterior angle once. Hence, the sum of the exterior angles is equal to 360°.

(b) The sum of each of the n pairs of interior angle and exterior angle is $180°$. Hence, the sum of all pairs is $n180°$. From (a), the sum of all exterior angles is $2 \times 180°$, and therefore the sum of all interior angles is $(n-2)180°$.

13. Find three ways of obtaining the basic (3,6,3,6) tessellation from other tessellations.

Solution. The basic (3,6,3,6) tessellation may be obtained from the basic (3,3,3,3,3,3) in two ways. First, we combine various sets of six equilateral triangles into regular hexagons. Second, we can cut each equilateral triangle into four equilateral triangles as shown in the figure below on the left. By merging equilateral triangles across six tiles, we obtain the basic (3,6,3,6) tessellation, as shown in the figure below on the right.

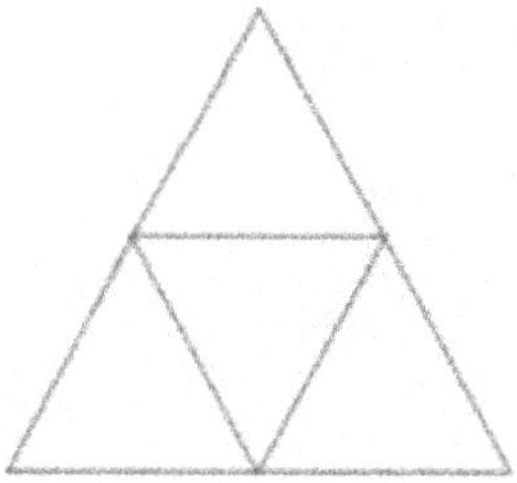

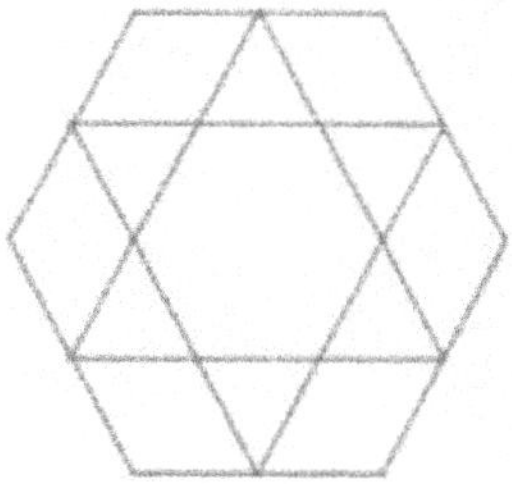

Finally, the basic (3,6,3,6) tessellation may be obtained from the (6,6,6) tessellation by cutting each regular hexagon into a regular hexagon and six congruent isosceles triangles with vertical angles 120°, as shown in the figure below on the left. When we merge isosceles triangles across three tiles, we obtain the basic (3,6,3,6) tessellation, as shown in the figure below on the right.

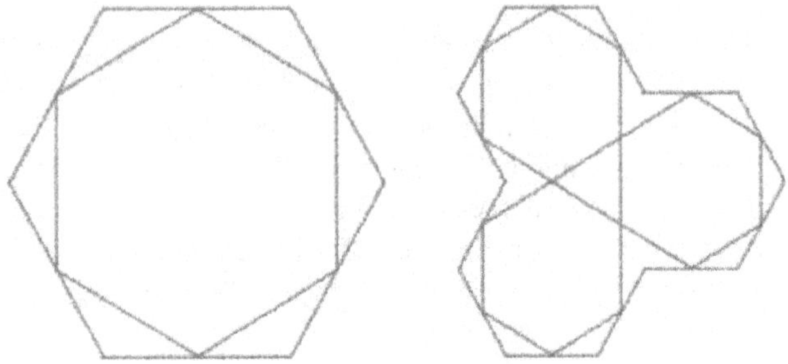

15. Find a way of obtaining the (3,4,6,4) tessellation from the (6,6,6) tessellation.

Solution. The (3,4,6,4) tessellation may be obtained from the (6,6,6) tessellation by cutting each regular hexagon into a regular hexagon, six congruent half-squares and six congruent kites with angles 120°, 90°, 60°, and 90°. as shown in the figure below on the left. When we merge half-squares across two tiles and kites across three tiles, we obtain the (3,4,6,4) tessellation, as shown in the figure below on the right.

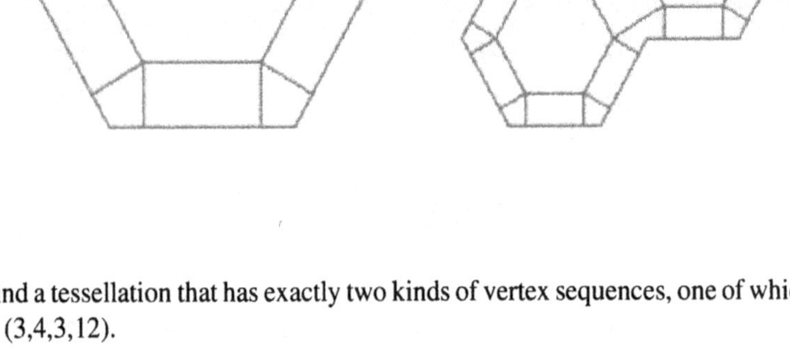

17. Find a tessellation that has exactly two kinds of vertex sequences, one of which is (3,4,3,12).

Solution. The tessellation in the figure below has two kinds of vertex sequences, (3,4,3,12) and (3,12,12).

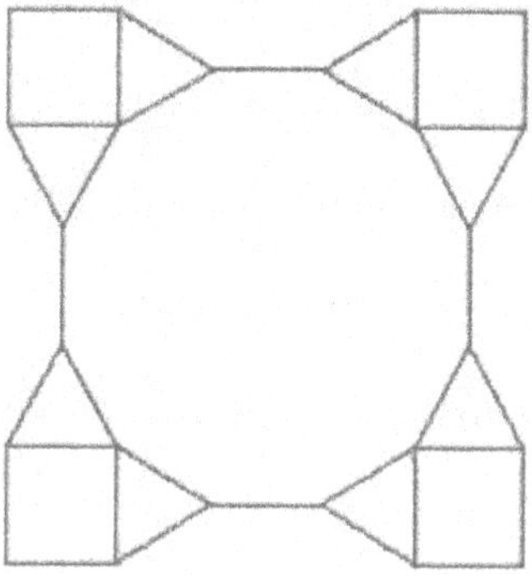

19. Find three tessellations that have exactly two kinds of vertex sequences, one of which is (3,4,4,6).

Solution. The tessellations in the figure below have two kinds of vertex sequences, one of which is (3,4,4,6). The others are (3,6,3,6), (3,6,3,6), and (3,4,6,4) respectively.

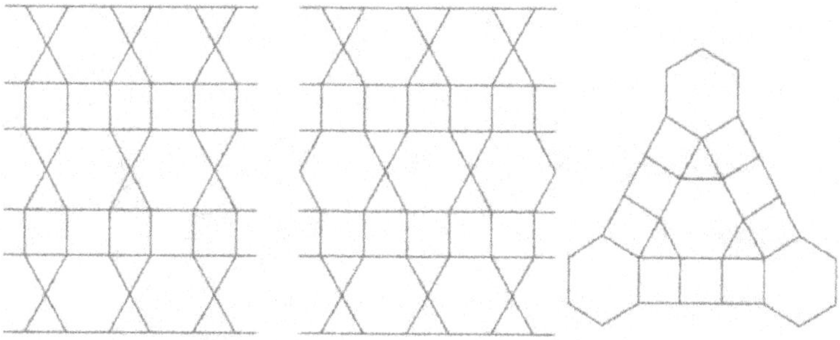

PART III

INVERSIVE AND PROJECTIVE GEOMETRIES

CHAPTER 13

INTRODUCTION TO INVERSIVE GEOMETRY

1. Suppose that P and Q are inverse points with respect to a circle with center S, that $SP = m$, and that the radius of the circle of inversion is n. Find SQ.

Solution. Since $SP \cdot SQ = n^2$, and $SP = m$, then $SQ = \dfrac{n^2}{m}$.

3. Given points P and Q with $PQ = 8$, draw all circles ω of radius 3 such that P and Q are inverses with respect to ω.

Solution. If O is the center of inversion as shown in the figure below, then

$$OP \cdot Oq = r^2.$$

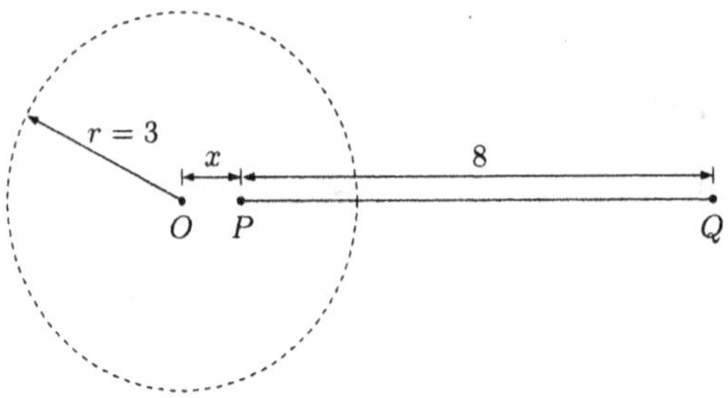

Letting x be the distance from O to P, we get $x(8 + x) = 3^2$. Solving for x, we get $x = 1$ and $x = -9$. This gives us two solutions. One solution has O on unit to the left of P as shown in the figure above.

The other solution has O nine units to the right of P. The two solutions are shown below.

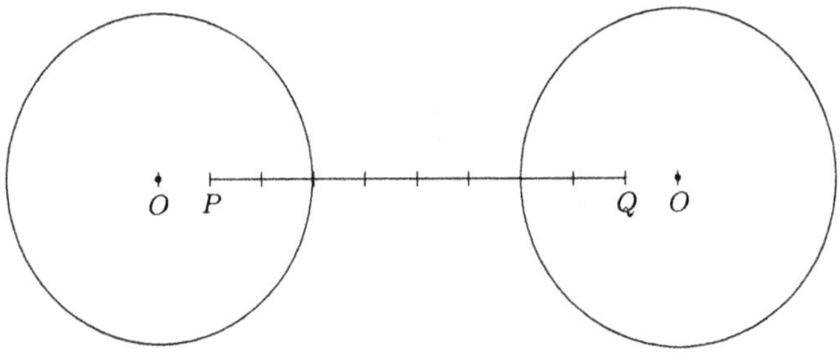

5. In the figure below, O is the center of the circle. The diameter ST is perpendicular to OP. PT intersects the circle at R, and SR intersects OP at Q. Prove that P and Q are inverses of each other with respect to the circle.

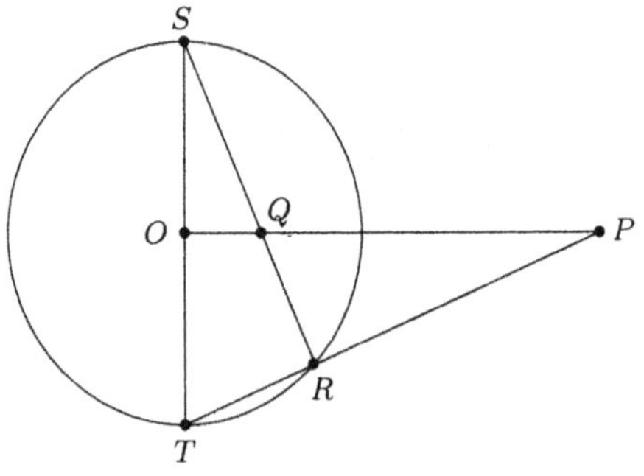

Solution. Note that $\angle SRT = 90°$ by Thales' Theorem. Also, since ST is perpendicular to OP, then $\angle QOS = 90°$. Since $\angle OQS = \angle PQR$, we have

$$\angle OSQ = 90° - \angle OQS = 90° - \angle PQR = \angle RPQ.$$

Thus, in $\triangle POT$ and $\triangle SOQ$ we have two angles congruent, so the triangles are similar, and

$$\frac{OP}{OT} = \frac{OS}{OQ}.$$

Therefore,

$$OP \cdot OQ = OT \cdot OS = r^2,$$

where r is the radius of the circle, and P and Q are inverses of each other with respect to the circle $\mathcal{C}(O, r)$.

7. Draw the figure obtained by inverting a square with respect to its circumcircle.

Solution. The solution consists of semicircles built on each side of the square, and is shown in the figure below.

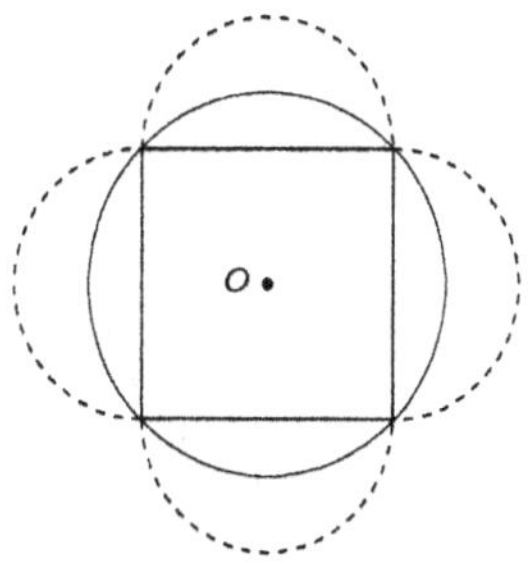

9. What is the inverse of a set of parallel lines?

Solution. The inverse is a set of circles tangent to each other at the center of inversion O.

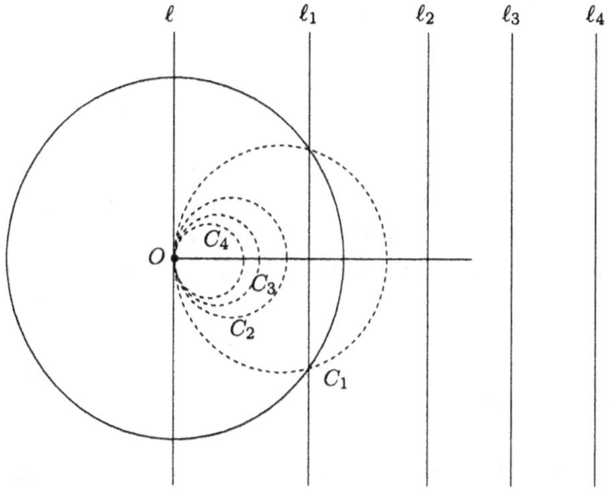

If the set of parallel lines includes a line ℓ through O, then ℓ is its own inverse and is also tangent to the circles at O.

11. Let P and Q have inverses P' and Q', respectively, under $I(O, r^2)$, with O between P and Q. Show that

$$P'Q' = \frac{PQ}{OP \cdot OQ} r^2.$$

This is called the ***distortion theorem***.

Solution.

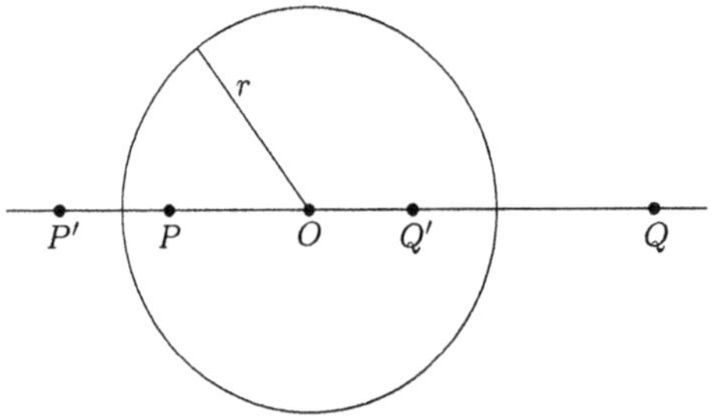

As in the figure above, we have

$$P'Q' = OP' + OQ' = \frac{r^2}{OP} + \frac{r^2}{OQ} = \frac{(OP + OQ)\, r^2}{OP \cdot OQ} = \frac{PQ}{OP \cdot OQ} r^2.$$

13. A circle and an intersecting line (nontangential) can be inverses to each other in two different ways. Illustrate this by showing how to find two circles of inversion α and β such that the line and the given circle are inverses of each other.

Solution. Let P and Q be the points where the given line meets the given circle. Let A and B be the points where the right bisector of PQ meets the given circle. These will be the centers of α and β. The circle α and β have radii AP and BP, respectively.

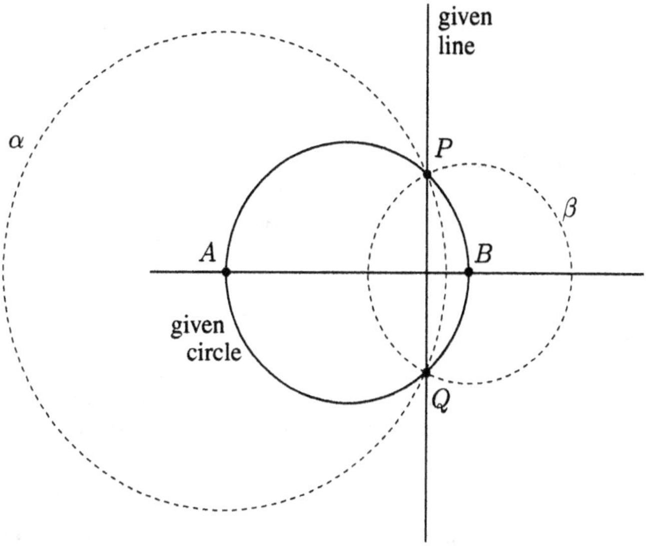

15. Given a triangle ABC with circumcenter O, let A', B', and C' be the images of the points A, B, and C under the inversion $I(O, r^2)$. Prove that

$$\triangle A'B'C' \sim \triangle ABC.$$

Solution. Let O be the circumcenter of $\triangle ABC$, and also the center of the circle of inversion ω, and let $\triangle A'B'C'$ be the image of $\triangle ABC$ as shown in the figure below.

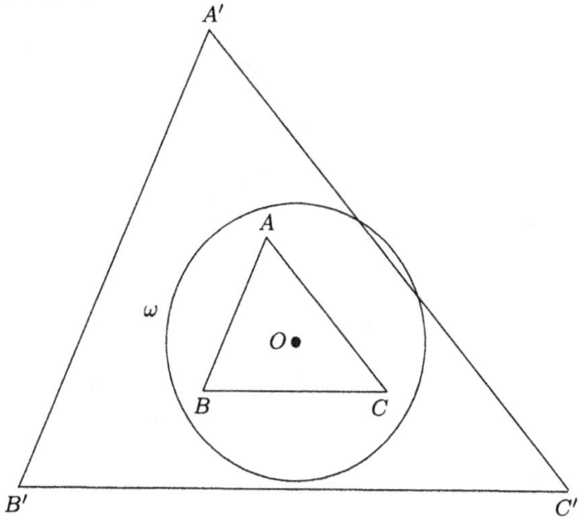

From the distortion theorem, we have

$$A'B' = \frac{AB}{OA \cdot OB} r^2, \quad B'C' = \frac{BC}{OB \cdot OC} r^2, \quad A'C' = \frac{AC}{OA \cdot OC} r^2.$$

Since O is the center of the circumcircle of $\triangle ABC$, we have

$$OA = OB = OC = R,$$

where R is the radius of the circumcircle, therefore

$$\frac{A'B'}{AB} = \frac{B'C'}{BC} = \frac{A'C'}{AC} = \left(\frac{r}{R}\right)^2.$$

Thus, the ratios are all equal, and by the **sss** similarity condition the triangles are similar.

Note that if the circle of inversion is the circumcircle, then $r = R$ and the triangles are congruent, since the vertices invert into themselves.

17. If PQ and RS are common tangents to two circles PAR and QAS, respectively, prove that the circles PAQ and RAS are tangent to each other.

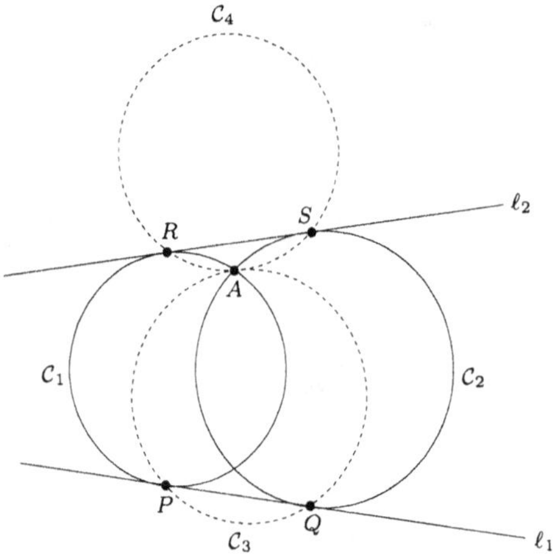

Solution. Let C_1 be the circle PAR and C_2 be the circle QAS, and let C_3 and C_4 be the circles PAQ and RAS, respectively. Also, let ℓ_1 and ℓ_2 be the common tangents PQ and RS, as in the figure above.

Since the angle between tangent circles and parallel lines is 0 and angles are invariant under inversion, in order to show that the circles PAQ and RAS are tangent, it suffices to show that $P'Q' \parallel R'S'$.

Inverting in a circle ω centered at A, we see that the inverse of the line PQ is a circle ℓ_1' through A, and the inverses of circles C_1 and C_2 are lines C_1' and C_2'. Since the line PQ is tangent to the circles C_1 and C_2 at P and Q, respectively, then ℓ_1' is tangent to C_1' and C_2' at the points P' and Q', respectively, as in the figure below. Similarly for the line RS and its inverse circle.

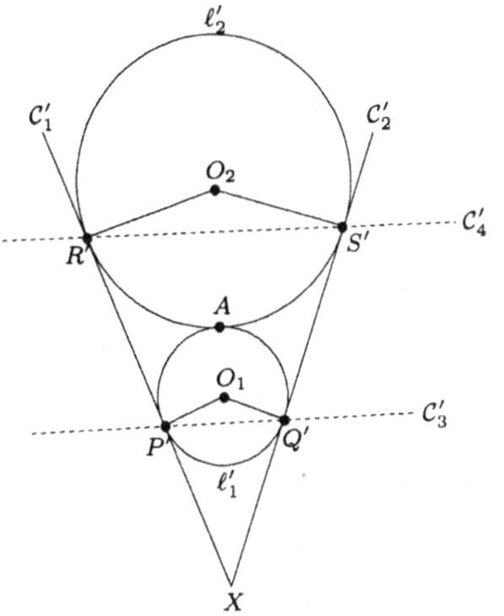

Since tangents from X have equal length, then $XP' = XQ'$ and $XR' = XS'$, so that $\triangle XP'Q'$ and $\triangle XR'S'$ are isosceles, and since $\angle P'XQ' = \angle R'XS'$, then $\triangle XP'Q' \sim \triangle XR'Q'$. Therefore, $\angle XP'Q' = \angle XR'S'$, so that $P'Q'$ is parallel to $R'S'$, and the circles PAQ and RAS are tangent at A.

19. Given a circle ω and two noninverse points P and Q inside ω, construct the circle through P and Q orthogonal to ω.

Solution. Construct P', the inverse of P with respect to ω, then the circle PQP' will be orthogonal to ω since it passes through a pair of inverse points.

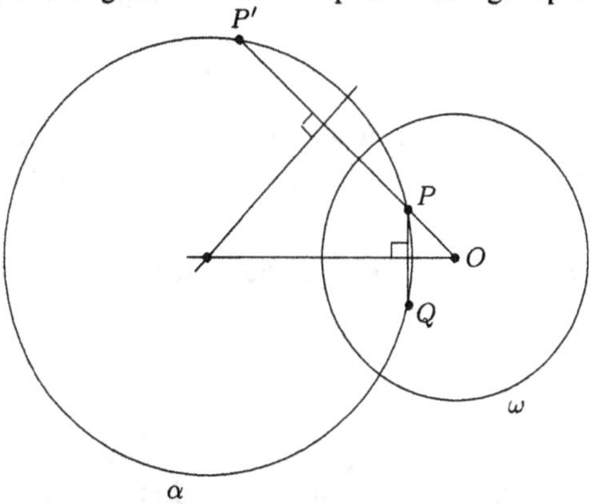

The points P, Q, and P' determine a unique circle α which is orthogonal to ω and which passes through the points P and Q. If P and Q are collinear with O (the center of ω), then the "circle" PQP' is just the line PQ.

21. Construct (using a straightedge and compass) a circle orthogonal to a given circle having within it one-third of the circumference of the given circle.

Solution.

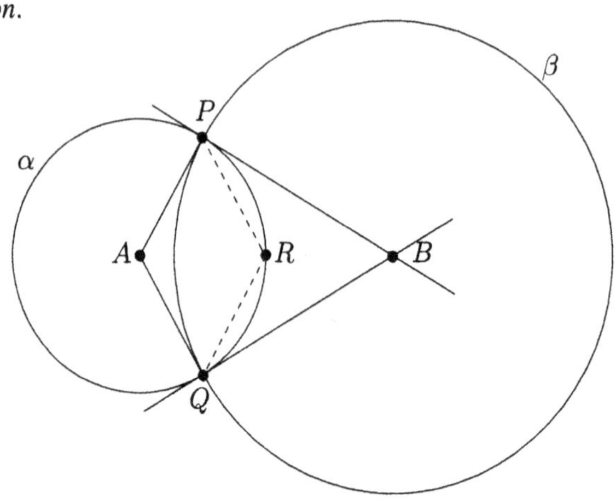

Let α denote the given circle, r denote the radius of α, and let A be its center.

(i) Choose a point P on α.

(ii) Strike off two points R and Q on α such that $PR = RQ = r$, then $\angle PAQ = 120°$ so that the arc PRQ is one-third of the circumference of the given circle.

(iii) Construct the perpendiculars to AP at P and AQ at Q, meeting at B, then $BP = BQ$ because BP and BQ are tangents to α.

(iv) With center B and radius BP draw the circle β, then β is the desired circle.

23. Let AC be a diameter of a given circle and chords AB and CD intersect (produced if necessary) in a point O. Prove that the circle OBD is orthogonal to the given circle.

Solution. In the figure on the following page, α is the given circle with AC as diameter, and the chords AB and CD intersect at O. We want to show that the circle β through O, B, and D, is orthogonal to α.

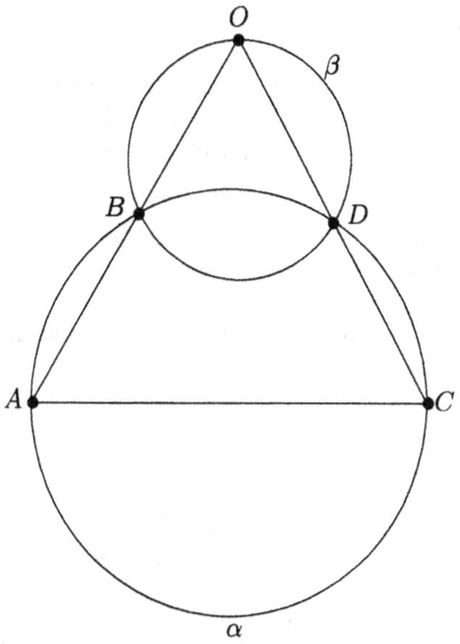

We invert in a circle ω centered at A, and note that since the line ODC does not pass through A, then the points O', D', and C' lie on a circle through A. Therefore, under the inversion, $AC'D'O'$ is a cyclic quadrilateral.

Since AC is a diameter of the circle α, then $B'C' \perp AC'$ as in the figure below.

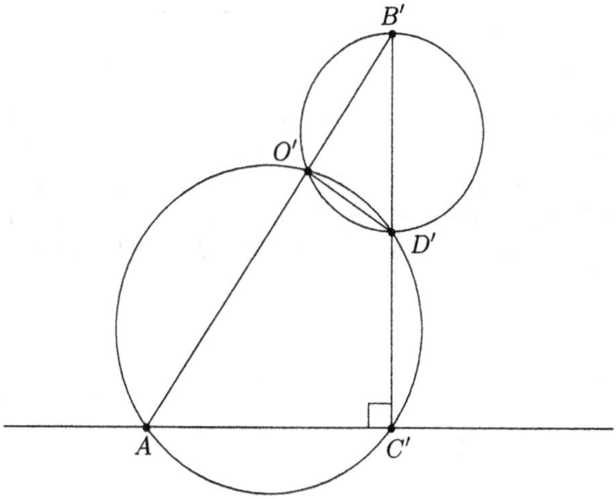

Therefore,

$$\angle AO'D' + \angle D'C'A = 180,$$

so that

$$\angle AO'D' + 90 = 180,$$

Hence, $\angle AO'D' = 90$, so that $B'O'D' = 90$.

Since $\triangle B'O'D'$ is inscribed in the circle, and $\angle B'O'D' = 90$, then $B'D'$ must be the diameter of circle $O'B'D'$. Thus, $B'D'$ is perpendicular to the circle $O'B'D'$, and since C' is collinear with B' and D', then the circle $O'B'D'$ is orthogonal to the circle $B'D'C'$.

Thus, inverting back, the circle OBD must be orthogonal to the circle BDC, since angles are preserved under inversion.

25. In the Arbelos Theorem, show that the points of contact of K_i and K_{i+1}, $i = 0, 1, \ldots$, lie on a circle.

Solution.

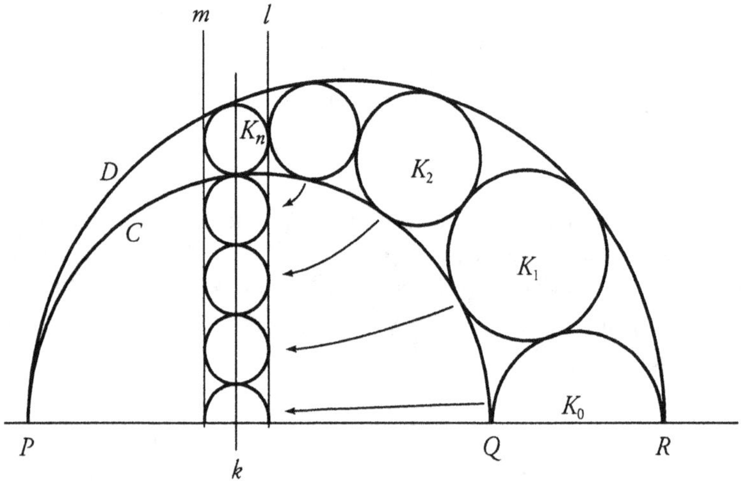

In the proof of the Arbelos Theorem, it is clear that the points of contact between the inverses K_i' and K_{i+1}' lie on a straight line k parallel to the lines l and m as shown in the above figure. Thus, k is the inverse of a circle k', through P. Since inversion preserves the points of contact, it follows that the points of contact between the original circles K_i and K_{i+1} must lie on k'.

CHAPTER 14

RECIPROCATION AND THE EXTENDED PLANE

1. Prove case (b) of statement (2) of Theorem 14.1.5.

   Solution. We have to show if O is the center and r the radius of the Circle of Apollonius for A, B, and k, the points A ad B are to the same side of the midpoint O of CD. In case (b) of statement (2), we may assume that the line AB is horizontal, that A is to the left of B, and that C is between A and B but D is not, and in fact, D is located to the right of B, as in the figure below.

   (b) 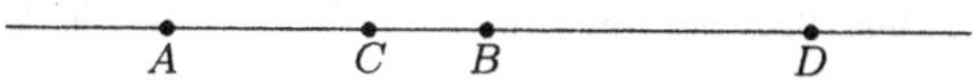

   For case (b), we have $CA < DA$ and since C and D are on the Circle of Apollonius, we also have
   $$\frac{BC}{CA} = \frac{BD}{DA},$$
   multiplying this equation by CA on the left and by DA on the right, we get
   $$BC = CA \cdot \frac{BC}{CA} < DA \cdot \frac{BD}{DA} = BD.$$
   Thus, the midpoint O of CD is to the right of B and hence to the right of both A and B.

Solutions Manual to Accompany Classical Geometry: Euclidean, Transformational, Inversive, and Projective, First Edition. By I. E. Leonard, J. E. Lewis, A. C. F. Liu, G. W. Tokarsky.

3. Prove Theorem 14.1.10: If A and B are inverse points for a circle ω, then ω is the Circle of Apollonius for A, B, and some positive number k.

Solution. Suppose that A and B are inverse points for a circle ω which has diameter CD as in the figure below.

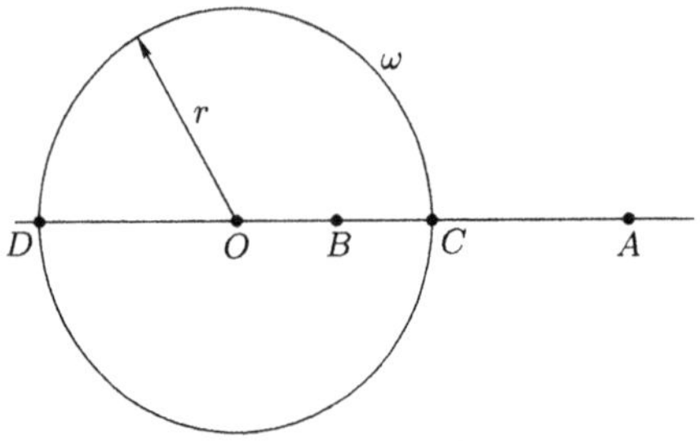

From Theorem 14.1.6, A and B are harmonic conjugates with respect to C and D, and if we let

$$k = \frac{AC}{CB} = \frac{AD}{DB},$$

then from Theorem 14.1.4,

$$\omega = \left\{ X \,\middle|\, \frac{AX}{XB} = k \right\}$$

is the Circle of Apollonius for A, B, and k.

5. Show that one of the angles between the polars of A and B is equal to $\angle AOB$, where O is the center of the reciprocating circle.

Solution. Let A and B be two points and let A' and B' be their inverses with respect to the circle ω centered at O. Let a and b be the polars of A and B, respectively, and let θ be the angle between a and b as shown in the figure below.

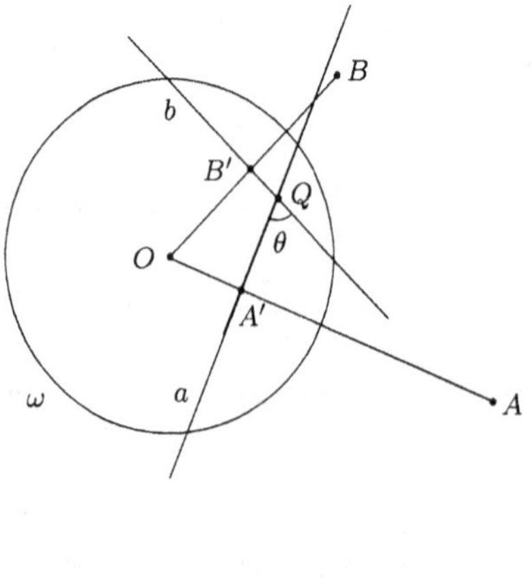

In the quadrilateral $OB'QA'$ we have opposite angles A' and B' equal to $90°$, so that

$$\angle A'OB' + \angle A'QB' = 180°.$$

From this we get

$$\angle A'OB' = 180 - \angle A'QB' = \theta,$$

so that $\angle AOB = \angle A'OB' = \theta$.

7. Use reciprocation to prove that given a triangle inscribed in a circle, then the points of intersection of the tangent lines at the vertices with the opposite sides are collinear.

   Solution. Suppose that $\triangle PQR$ is inscribed in a circle as shown in the figure below. Let AP and AR be the tangents to the circle at P and R, respectively, intersecting at A, then $a = PR$ is the polar of A.

   Similarly, let BP and BQ be the tangents to the circle at P and Q, respectively, intersecting at B, then $b = PQ$ is the polar of B.

   Similarly, let CQ and CR be the tangents to the circle at Q and R, respectively, intersecting at C, then $c = QR$ is the polar of C.

   Finally, let

$$b \cap AC = N,$$
$$c \cap AB = L,$$
$$a \cap BC = M,$$

   it suffices to show that L, M, and N lie on the polar of O, the Gergonne point of $\triangle PQR$.

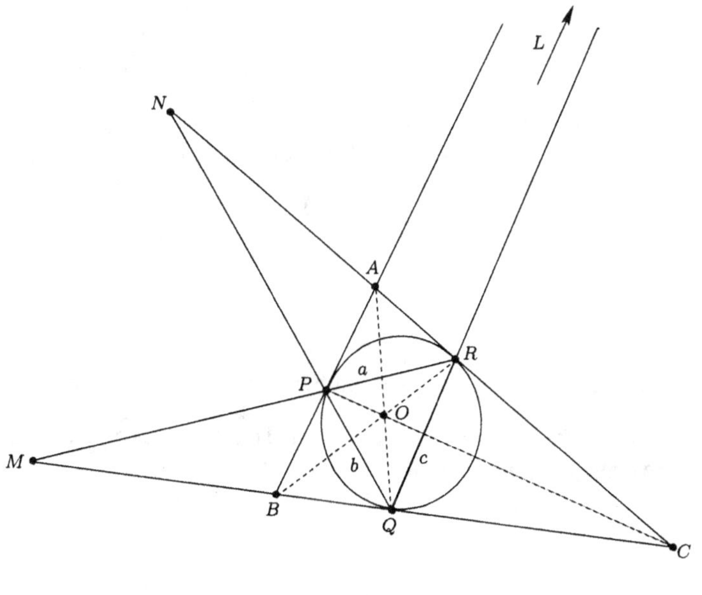

Since N lies on the polar of B, then B must lie on the polar of N, so that N is the pole of BR.

Similarly, M is the pole of QA, and L is the pole of CP, so that L, M, N lie on the polar of O (O lies on the polars of L, M, and N). Therefore, L, M, and N are collinear.

Alternate Solution. Suppose that the triangle $\triangle ABC$ is inscribed in the circle ω as shown in the figure below, and the tangents and the sides intersect at D, E, and F as shown.

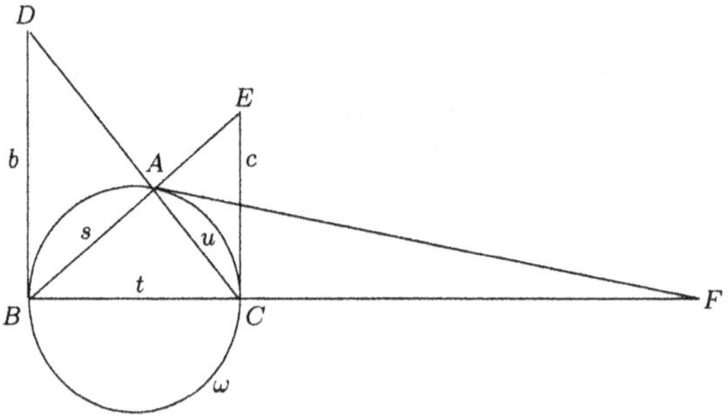

Suppose that the figure on the previous page reciprocates in ω to give the figure below.

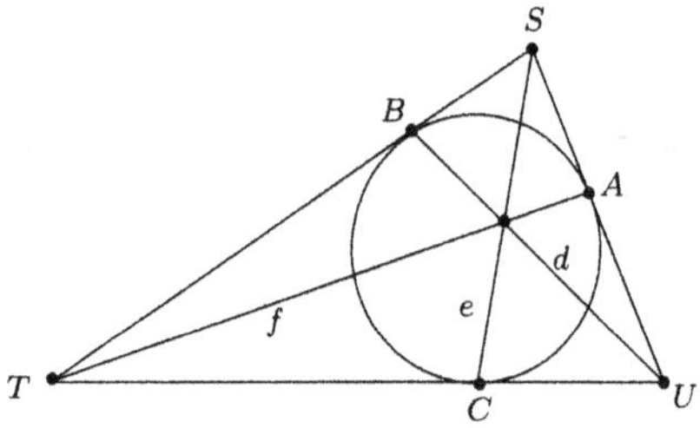

We need to prove that f, e, and d are concurrent in this diagram. However, they meet at the Gergonne point, since by Ceva's theorem, the lines are concurrent:

$$\begin{aligned}\frac{\overline{SB}}{\overline{BT}} \cdot \frac{\overline{TC}}{\overline{CU}} \cdot \frac{\overline{UA}}{\overline{AS}} &= \frac{SB}{BT} \cdot \frac{TC}{CU} \cdot \frac{UA}{AS} \\ &= 1 \cdot 1 \cdot 1 \\ &= 1,\end{aligned}$$

since all the ratios are internal and $SB = SA$, $TB = TC$, and $UA = UC$.

9. If two circles are orthogonal, prove that the extremities of any diameter of one are conjugate points for the other.

Solution. In the figure, P and Q are the extremities of a diameter in circle α.

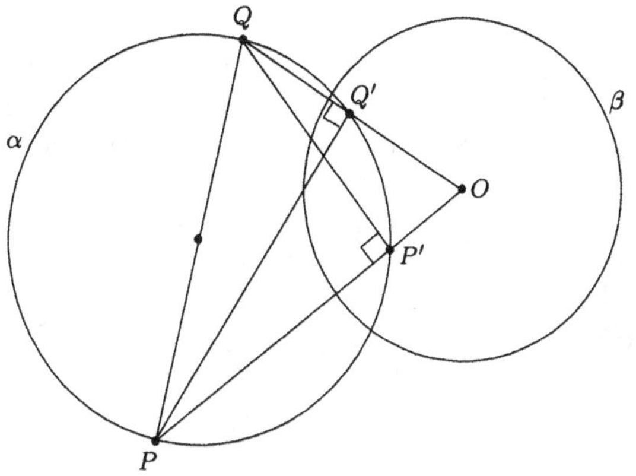

Since the circles α and β are orthogonal, from Problem 14.4, the line OQ cuts α in inverse points with respect to β. Similarly, the line OP cuts α in inverse points with respect to β.

Now consider the angle $\angle QQ'P$ and $\angle PP'Q$, since PQ is a diameter of α, then $\triangle PQQ'$ and $\triangle PQP'$ are inscribed in a semicircle, and therefore

$$\angle QQ'P = 90 = \angle PP'Q.$$

Thus, Q' lies on the segment OQ which is perpendicular to PQ' at Q', so that the line PQ' is the polar of Q with respect to β. Similarly, P' lies on the segment OP which is perpendicular to QP' at P', so that the line QP' is the polar of P with respect to β. Therefore, P and Q are conjugate points for β.

11. Given r and ϵ, $0 < \epsilon < 1$, then the equation

$$\frac{FX}{\text{dist}(X, d)} = \epsilon$$

in Cartesian coordinates becomes

$$\left(x - \frac{r\epsilon^2}{1-\epsilon^2}\right)^2 + \frac{y^2}{1-\epsilon^2} = \left(\frac{r\epsilon}{1-\epsilon^2}\right)^2.$$

Show that given positive numbers a and b, there are suitable values for r and ϵ (in terms of a and b) so that the Cartesian equation above becomes

$$\frac{(x-f)^2}{a^2} + \frac{y^2}{b^2} = 1.$$

Solution. In the focus-directrix definition of a conic

$$\frac{FX}{\text{dist}(X, d)} = \epsilon,$$

d is a fixed line, F is a fixed point in the plane not on the line, dist(X, d) is the perpendicular distance from the point X to the line d, and ϵ is a fixed positve constant.

To obtain the Cartesian equations of the conic, we assume that the focus F is at the point $(0,0)$, d is a vertical line perpendicular to the x-axis through the point $(-r, 0)$. The Cartesian equations of the conic are then

$$\left(x - \frac{r\epsilon^2}{1-\epsilon^2}\right)^2 + \frac{y^2}{1-\epsilon^2} = \left(\frac{r\epsilon}{1-\epsilon^2}\right)^2.$$

This is of the form

$$\frac{(x-f)^2}{a^2} + \frac{y^2}{b^2} = 1.$$

provided

$$a^2 = \left(\frac{r\epsilon}{1-\epsilon^2}\right)^2$$

$$b^2 = (1-\epsilon^2)\left(\frac{r\epsilon}{1-\epsilon^2}\right)^2.$$

Dividing the second of these two equations by the first, we get

$$\frac{b^2}{a^2} = \frac{(1-\epsilon^2)\left(\frac{r\epsilon}{1-\epsilon^2}\right)^2}{\left(\frac{r\epsilon}{1-\epsilon^2}\right)^2},$$

that is,

$$\frac{b^2}{a^2} = 1 - \epsilon^2,$$

and solving for ϵ we get

$$\epsilon = \sqrt{1 - \frac{b^2}{a^2}}.$$

Substituting this into the first equation, and solving for r we get

$$r = \frac{b^2}{\sqrt{a^2 - b^2}}.$$

13. Find the Cartesian equations of the following conic sections:

    (a) foci: $(\pm 8, 0)$, $e = 0.2$.

    (b) foci: $(\pm 4, 0)$, directrix: $x = \dfrac{16}{3}$.

Solution.

(a) We have

$$c = 8 \qquad \text{and} \qquad e = 0.2 = \frac{c}{a},$$

so that

$$a = \frac{8}{.2} = 40.$$

Since $e < 1$, we have an ellipse, and

$$b^2 = a^2 - c^2 = 40^2 - 8^2 = 1536.$$

The equation is

$$\frac{x^2}{1600} + \frac{y^2}{1536} = 1.$$

(b) This is either an ellipse or an hyperbola, and since the directrix is not between the foci, it must be an ellipse.

We have

$$e = \frac{c}{a} \qquad \text{and} \qquad x = \frac{a}{e}$$

so that

$$a^2 = cx, \qquad \text{and} \qquad a^2 = \frac{64}{3}.$$

Thus,

$$b^2 = a^2 - c^2 = \frac{64}{3} - 16 = \frac{16}{3},$$

and the equation is

$$\frac{x^2}{64/3} + \frac{y^2}{16/3} = 1.$$

CHAPTER 15

CROSS RATIOS

1. Given $(AB, CD) = k$, find (BC, AD) and (BD, CA).

   Solution. We have
   $$(AB, CD) = k,$$
   which implies
   $$(BA, CD) = \frac{1}{k},$$
   which in turn implies
   $$(BC, AD) = 1 - \frac{1}{k}.$$
   Also,
   $$(AB, CD) = k,$$
   which implies
   $$(DB, AC) = 1 - k,$$
   which in turn implies
   $$(BD, CA) = \frac{1}{1 - k}.$$

3. Using the definition of the cross ratio, show that $(AB, CD) = (CD, AB)$.

Solution. We have

$$(AB, CD) = \frac{\overline{AC}/\overline{CB}}{\overline{AD}/\overline{DB}} = \frac{\overline{AC} \cdot \overline{DB}}{\overline{CB} \cdot \overline{AD}},$$

so that

$$(CD, AB) = \frac{\overline{CA}/\overline{AD}}{\overline{CB}/\overline{BD}} = \frac{\overline{CA} \cdot \overline{BD}}{\overline{AD} \cdot \overline{CB}} = \frac{(-\overline{AC}) \cdot (-\overline{DB})}{\overline{AD} \cdot \overline{CB}} = \frac{\overline{AC} \cdot \overline{DB}}{\overline{CB} \cdot \overline{AD}}.$$

5. Find x if $(AB, CD) = (BA, CD) = x$.

Solution. If

$$(AB, CD) = x$$

then this implies that

$$(BA, CD) = \frac{1}{x},$$

so that

$$x = \frac{1}{x}.$$

Therefore, $x^2 = 1$, so that either $x = +1$ or $x = -1$. Since a cross ratio can never be $+1$, we must have $x = -1$.

7. If P, Q, and R are the respective feet of the altitudes on the sides of BC, CA, and AB of $\triangle ABC$, show that

$$P(QR, AB) = -1.$$

We give three different solutions.

Solution 1. Let the line through P and Q intersect the side AB at the point X, as in the figure below.

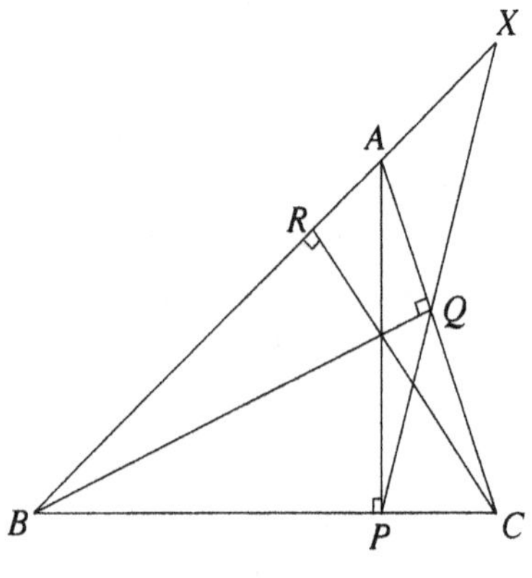

We have

$$P(QR, AB) = (XR, AB) = \frac{\overline{XA}/\overline{AR}}{\overline{XB}/\overline{BR}} = \frac{\overline{XA}}{\overline{AR}} \cdot \frac{\overline{RB}}{\overline{BX}},$$

and we need only show that

$$\frac{\overline{XA}}{\overline{AR}} \cdot \frac{\overline{RB}}{\overline{BX}} = -1.$$

From Ceva's Theorem, we have

$$\frac{\overline{CQ}}{\overline{QA}} \cdot \frac{\overline{AR}}{\overline{RB}} \cdot \frac{\overline{BP}}{\overline{PC}} = 1,$$

while from Menelaus's Theorem, we have

$$\frac{\overline{AX}}{\overline{XB}} \cdot \frac{\overline{BP}}{\overline{PC}} \cdot \frac{\overline{CQ}}{\overline{QA}} = -1.$$

Dividing the first equation by the second, we obtain

$$\frac{\overline{AR}}{\overline{RB}} \cdot \frac{\overline{XB}}{\overline{AX}} = -1,$$

that is,

$$\frac{\overline{XA}}{\overline{AR}} \cdot \frac{\overline{RB}}{\overline{BX}} = -1,$$

so that

$$(XR, AB) = P(QR, AB) = -1.$$

Solution 2. Note that the second solution requires knowledge about the relationship between harmonic conjugates and cross ratios.

From Theorem 15.1.7, we have

$$P(QR, AB) = (SR, AB) = -1,$$

since R and S are harmonic conjugates with respect to A and B, that is, AC, BC, AP, BQ form a complete quadrilateral.

Solution 3. Referring to the figure on the following page, we have

$$(SR, AB) = (SX, PQ) = (SR, BA) = \frac{1}{(SR, AB)},$$

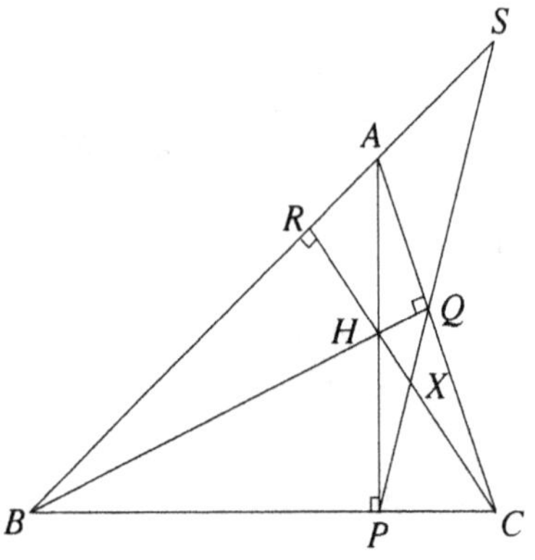

Where in the first equality the pole is at H, in the second equality the pole is at C, and in the third equality we have interchanged the last pair.

Therefore $(SR, AB)^2 = 1$, so that $(SR, AB) = \pm 1$, and since S and R are separated by A and B, then $(SR, AB) = -1$.

9. Prove the following: If PA, PB, PC, PD and QA, QB, QC, QD are two pencils of lines, and if $P(AB, CD) = Q(AB, CD)$ and A is on PQ, then B, C, and D are collinear.

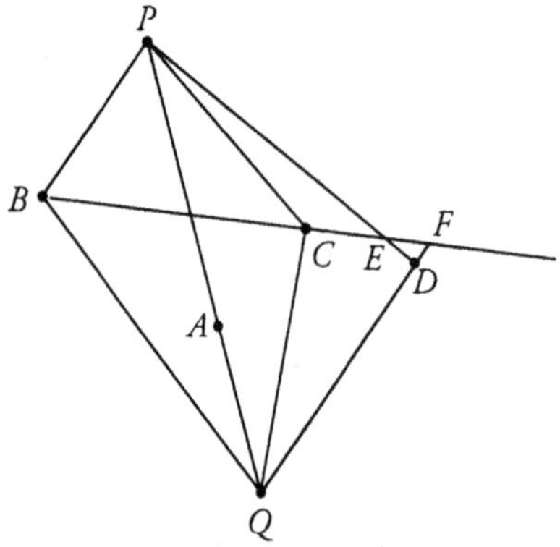

Solution. Let E and F be the points where PD and QD intersect the line BC, then

$$P(AB, CD) = P(AB, CE) = (AB, CE)$$

and

$$Q(AB, CD) = Q(AB, CE) = (AB, CF).$$

Since $P(AB, CD) = Q(AB, CD)$, we have $(AB, CE) = (AB, CF)$ which implies that $E = F$. Thus PE and QF intersect at the point E. However, PE and QF intersect at D, so $E = F = D$, and therefore D is on the line BC

11. Given a variable triangle $\triangle ABC$ whose sides BC, CA, and AB pass through fixed points P, Q, and R, respectively, then if the vertices B and C move along

given lines through a point O collinear with Q and R, find the locus of the vertex A.

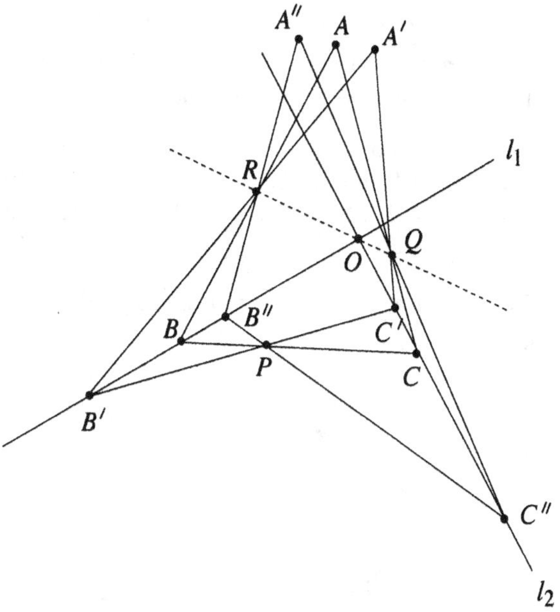

Solution. From the figure we have

$$R(OA, A'A'') = R(OB, B'B''),$$

since these arise from the same pencil of lines.

Also, since O, B'', B, B', are collinear, then

$$R(OB, B'B'') = (OB, B'B'') = P(OB, B'B'').$$

Again, from the figure we have

$$P(OB, B'B'') = P(OC, C'C''),$$

since these arise from the same pencil of lines.

Also, since O, C, C', C'' are collinear, then

$$P(OC, C'C'') = (OC, C'C'') = Q(OC, C'C'').$$

Again, from the figure we have

$$Q(OC, C'C'') = Q(OA, A'A''),$$

since these arise from the same pencil of lines.

Therefore,

$$R(OA, A'A'') = Q(OA, A'A''),$$

so that R, O, Q, are collinear, and Lemma 15.2.4 implies that A', A, A'', are collinear. Thus, the locus of A is a straight line.

13. The bisector of angle A of $\triangle ABC$ intersects the opposite side at the point T. The points U and V are the feet of the perpendiculars from B and C, respectively, to the line AT. Show that U and V divide AT harmonically; that is, that $(AT, UV) = -1$.

Solution. Let h_1 and h_2 be the length of the perpendiculars BU and CV, respectively.

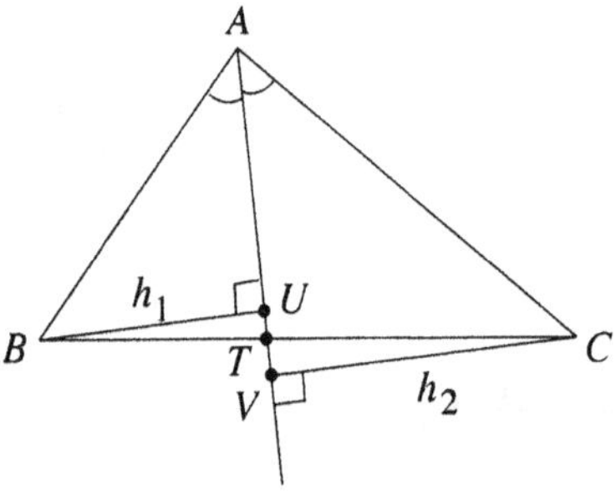

By the **AA** similarity theorem, $\triangle ABU \sim \triangle ACV$, so that

$$\frac{AU}{AV} = \frac{h_1}{h_2}.$$

Also, by the **AA** similarity theorem, $\triangle BUT \sim \triangle CVT$, so that

$$\frac{TU}{TV} = \frac{h_1}{h_2}.$$

Therefore, since U and V separate A and T, we have

$$\begin{aligned}(AT, UV) = \frac{\overline{AU}/\overline{UT}}{\overline{AV}/\overline{VT}} &= -\frac{AU/UT}{AV/VT} \\ &= -\frac{AU}{AV} \cdot \frac{TV}{TU} \\ &= -\frac{h_1}{h_2} \cdot \frac{h_2}{h_1} \\ &= -1,\end{aligned}$$

so that $(AT, UV) = -1$.

15. Prove the second part of Desargues' Theorem using cross ratios; that is, show that coaxial triangles are copolar.

Solution.

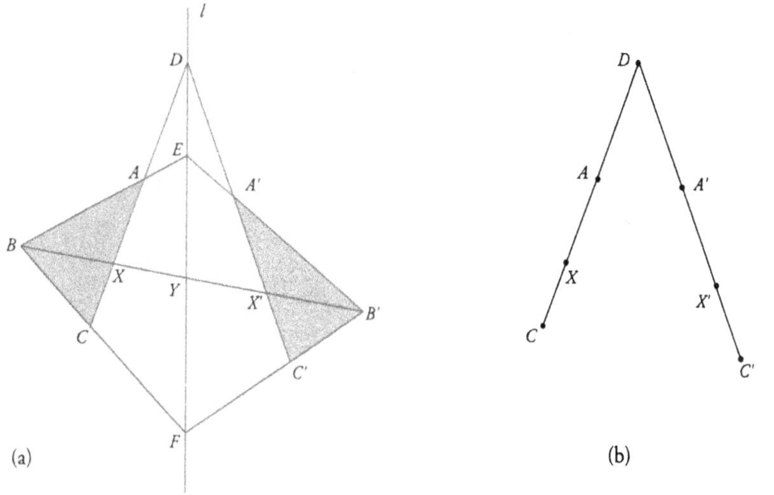

Triangles ABC and $A'B'C'$ have corresponding sides intersecting at D, E, F, and since the triangles are coaxial, D, E, and F are collinear as in figure (a) above.

Now let X, X', and Y be defined as follows

$$X = AC \cap BB', \qquad X' = A'C' \cap BB', \qquad Y = DF \cap BB'.$$

We will show that AA', CC', and XX' are concurrent, and since the lines XX' and BB' are identical, this will prove the theorem. In order to show this, it suffices to show that

$$(DA, XC) = (DA', X'C'),$$

since by Lemma 15.2.4, this implies that AA', CC', and XX' are concurrent as in figure (b) above.

Referring to figure (a),

$$\begin{aligned}
(DA, XC) &= B(DA, XC) \quad (DAXC \text{ is a transversal for the pencil at } B)\\
&= B(DE, YF) \quad (DEYF \text{ is a transversal for the pencil at } B)\\
&= (DE, YF)\\
&= B'(DE, YF) \quad (DEYF \text{ is a transversal for the pencil at } B')\\
&= B'(DA', X'C) \; (DA'X'C' \text{ is a transversal for the pencil at } B')\\
&= (DA', X'C).
\end{aligned}$$

This finishes the proof.

An outline of an alternate proof is given below.

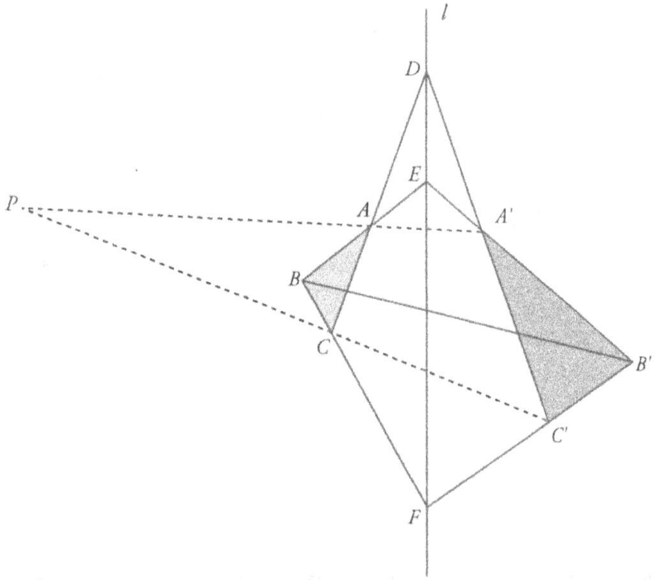

Let $P = AA' \cap CC'$, and note that $\triangle AEA'$ and $\triangle CFC'$ are copolar from the point D. The proof from the first part of the theorem shows that B, B', and P are collinear.

CHAPTER 16

INTRODUCTION TO PROJECTIVE GEOMETRY

1. Prove that the construction of the midpoint in Example 16.1.2 works by using the fact that the lines l, m, and CO are the diagonals of a complete quadrilateral.

   Solution. As in the figure below, consider l, m, and CO to be the diagonals of the complete quadrilateral with sides AC, AD, BC, and BE, then CO and m divide AB harmonically at F and I.

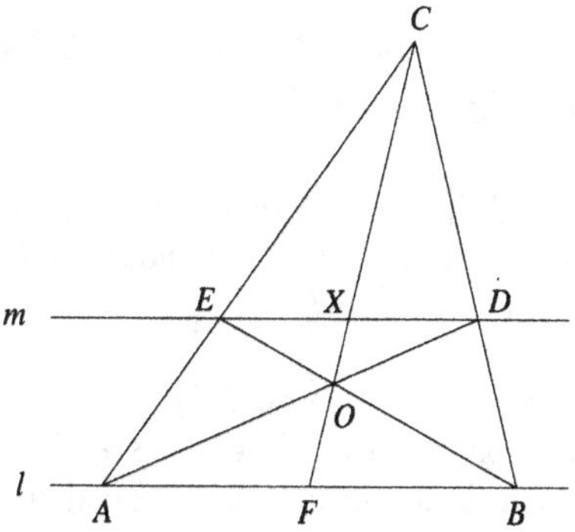

Solutions Manual to Accompany Classical Geometry: Euclidean, Transformational, Inversive, and Projective, First Edition. By I. E. Leonard, J. E. Lewis, A. C. F. Liu, G. W. Tokarsky.

Since I is an ideal point,

$$\frac{\overline{AI}}{\overline{IB}} = -1,$$

and since F is the harmonic conjugate of I, we must have

$$\frac{\overline{AF}}{\overline{FB}} = +1,$$

and this implies that $AF = FB$.

3. Using a straightedge alone, is it possible to construct a right angle? Explain.

Solution. Suppose that we could construct a right angle with a straightedge alone, we will show that this implies it is possible to find the center of a circle with a straightedge alone.

Given a circle ω without its center, at a point A on the circle construct a right angle whose arms intersect the circle at P and Q. From the converse of Thales' Theorem, the segment PQ is a diameter of the circle.

Now repeat this process at a point B on ω to find another diameter RS of the circle, then the intersection of PQ and RS gives the center O of the circle, as in the figure below.

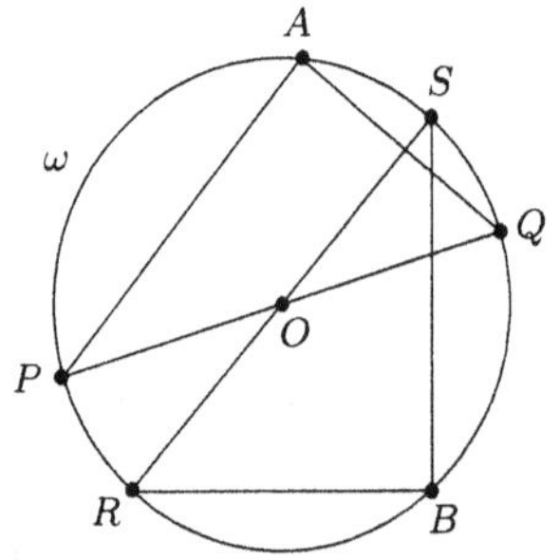

However, we showed Theorem 16.1.6 that it is impossible to construct the center of a circle with a straightedge alone.

Alternate Solution. We will show that if it is possible to construct a right angle with a straightedge alone, then we can bisect a line segment using a straightedge alone.

Given a line segment AB, construct right angles at the endpoints A and B, each with one of the arms along AB. Let ℓ and m be the lines through A and B, respectively, perpendicular to AB.

Construct points P on ℓ and Q on m such the the segment PQ intersects AB at an interior point. Next construct right angles at P and Q with arms along ℓ and m, repectively, and the other arms intersecting m and ℓ at R and S, respectively.

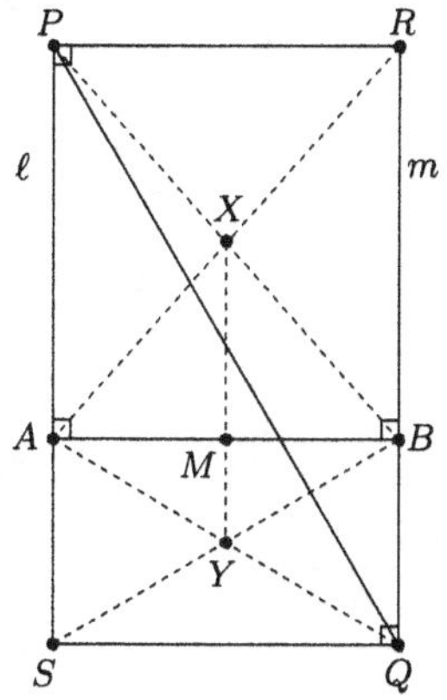

As in the figure above, $PABR$ and $SABQ$ are rectangles, so their diagonals bisect each other at X and Y, respectively. Hence, the line segment XY intersects AB at the midpoint M. However, this contradicts Theorem 16.1.1.

5. Given a circle ω without its center, and given a point P outside ω, construct the tangents to ω from P.

   Solution. Construct the polar p of P as in the proof of Theorem 16.1.9. Let S and T be the intersection points of p with ω, as in the figure below.

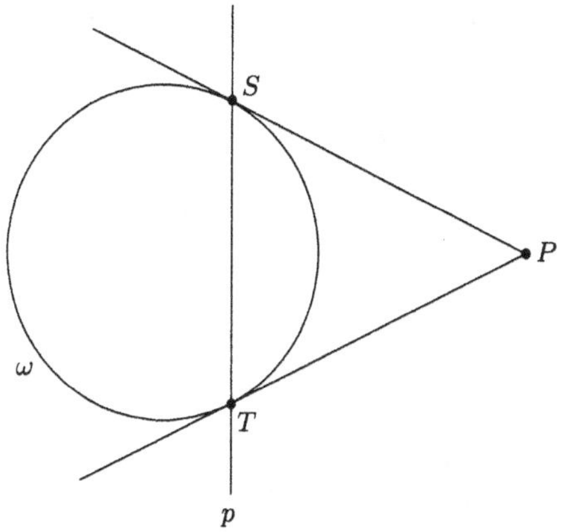

   Since S lies on the polar of P, then P lies on the polar of S, but the polar of S is the tangent at S. Thus, the lines PS and PT are the tangents to ω from P.

7. Given a circle ω without its center, and given a point P on ω, construct the tangent to ω at P.

   Solution. This amounts to constructing the polar p of P. As in the figure below, draw any chord PR and let V_1 be a point on PR. Now construct v_1, the polar of V_1, as in Problem 16.6.

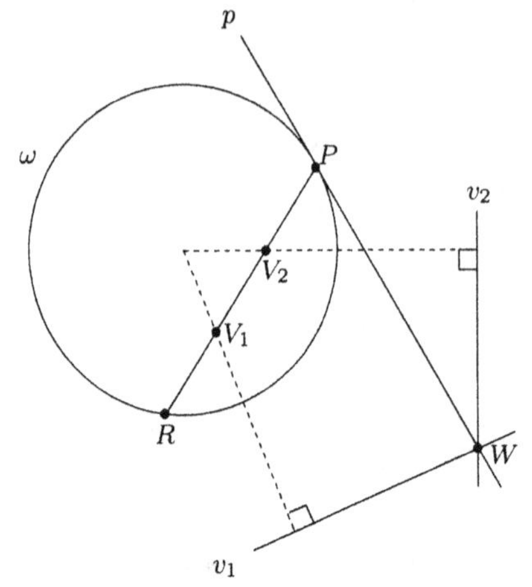

Now repeat this construction with another point V_2 on PR, leading to its polar v_2. Since P, V_1, and V_2 are collinear, their polars p, v_1, and v_2 are concurrent, so that $v_1 \cap v_2$ is on p. Let $W = v_1 \cap v_2$, the line PW is the desired polar p.

9. Given that π is a central perspectivity from l to l', show that the information that π maps A, B to A, B', respectively, is **not** sufficient to determine π uniquely.

Solution. Since A gets projected to itself, this means that $A = l \cap l'$. Let O_1 and O_2 be two points on BB', as shown in the figure.

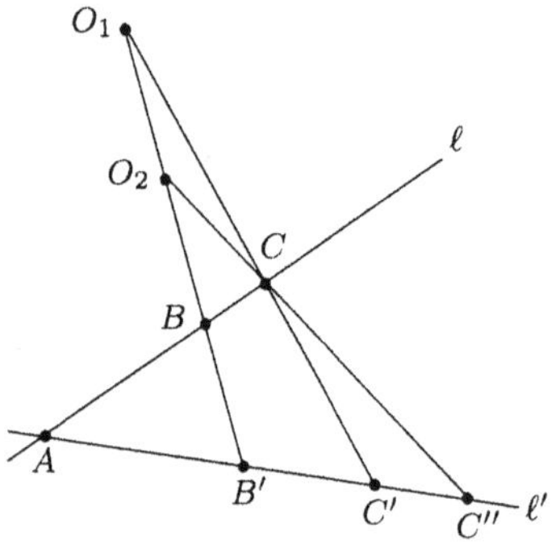

Let π_1 be the perspectivity centered at O_1 and π_2 be the perspectivity centered at O_2. Both perspectivities map $\{A, B\}$ to $\{A, B'\}$ but $\pi_1(C) \neq \pi_2(C)$ so the perspectivity is not determined uniquely.

11. In the figure on the following page, there is a line perspectivity from the pencil at A to the pencil at B that takes l to l' and m to m'. Using only a straightedge, construct the image of n under this perspectivity.

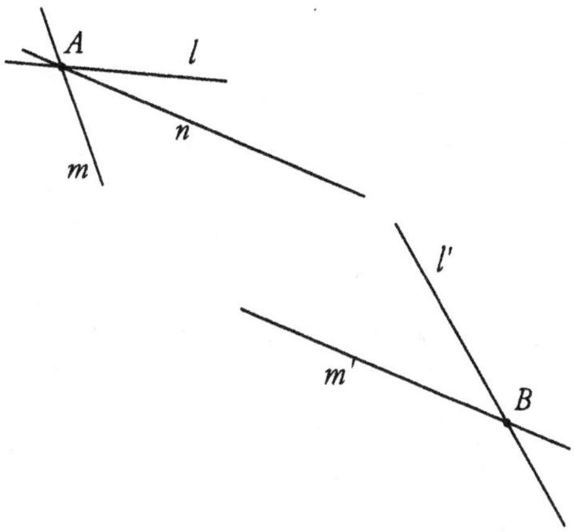

Solution. Let $P = l \cap l'$ and $Q = m \cap m'$, extending l, l', m, and m' as necessary, then PQ is the axis of perspectivity.

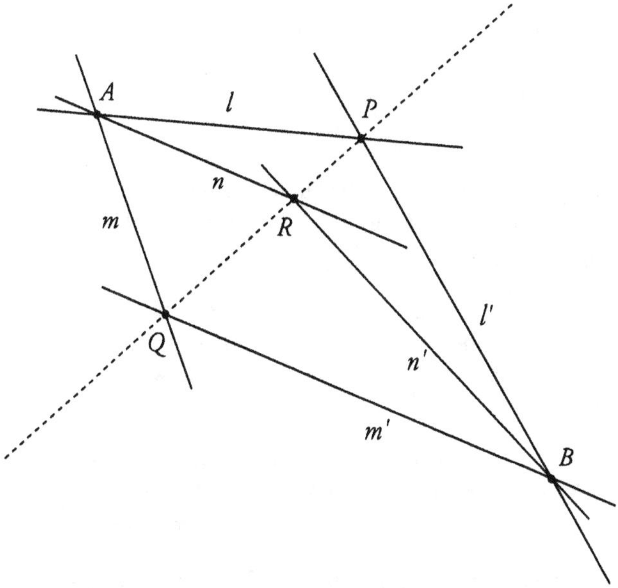

Now construct $R = PQ \cap n$, then the image of n under the perspectivity is $n' = \overline{BR}$.

13. Give an example of a projectivity from m onto m, the same line, with two distinct fixed points, but which is not a perspectivity.

Solution. Given A, B on the line m, as in the figure on the following page, draw m', m'' so that m, m', and m'' are not concurrent.

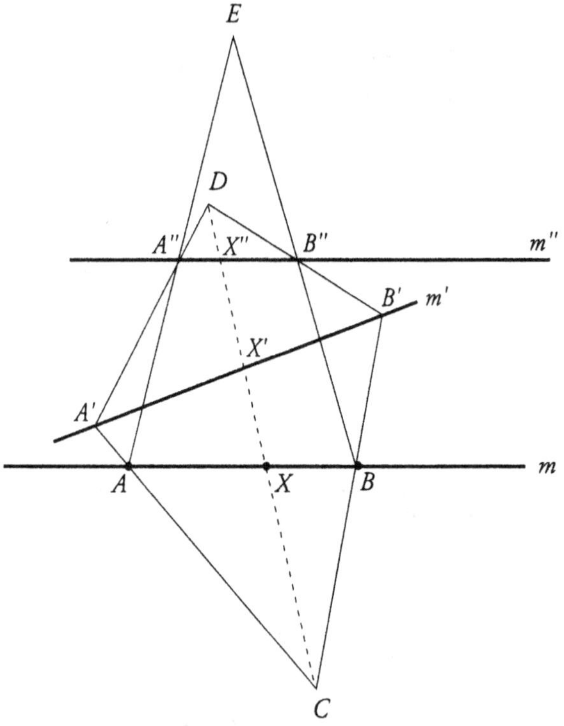

Use a perspectivity from a center C to map m to m', and a perspectivity from another center D to map m' to m''. The first takes $\{A, B\}$ to $\{A', B'\}$. The second takes $\{A', B'\}$ to $\{A'', B''\}$.

Let $E = AA'' \cap BB''$, and use E as a center of perspectivity that takes m'' to m. The combination of the three perspectivities is a projectivity from m to itself. The projectivity leaves A and B fixed, and we will show that no other point is fixed.

If any point other than A or B is fixed, then three points are fixed, and the projectivity would be a perspectivity from m to m and all points would be fixed. Thus, we only have to show that at least one point is not fixed.

We will show that if $X = CD \cap m$, then X is not fixed.

- The first perspectivity takes X to X' where $X' = CD \cap m'$.
- The second perspectivity takes X' to X'' where $X'' = CD \cap m''$.
- The third perspectivity takes X'' to some point X''' on m.

If $X''' = X$, then all of the points C, D, E, X, X', and X'' would be collinear. This would mean that triangles $AA'A''$ and $BB'B''$ are coaxial since

$$AA' \cap BB' = C, \qquad AA'' \cap BB'' = E, \qquad A'A'' \cap B'B'' = D.$$

By Desargue's Theorem, the triangles would have to be copolar, that is, the lines $AB = m$, $A'B' = m'$, and $A''B'' = m''$ would be concurrent However, we know that they are not concurrent. This shows that X is not fixed.

15. Let $PQRS$ be a complete quadrangle, and let

$$A = PQ \cap RS, \quad B = PR \cap SQ, \quad C = PS \cap QR, \quad D = AB \cap PS.$$

Show that $(PS, DC) = -1$ by projecting the line AC to infinity.

Solution. As in the figure below,

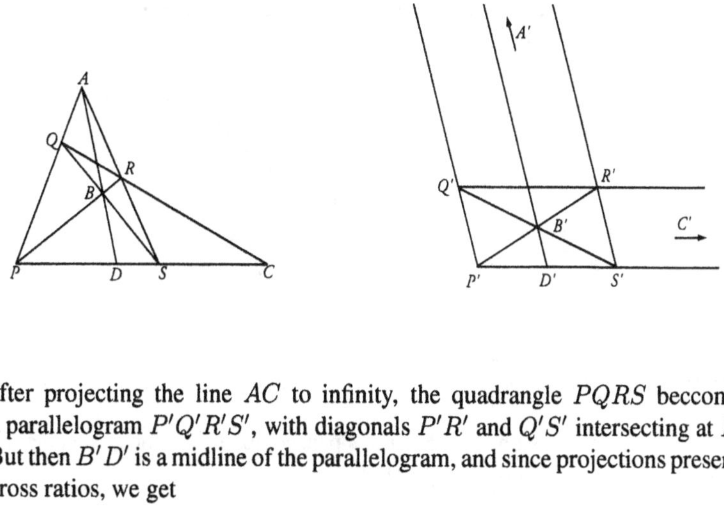

after projecting the line AC to infinity, the quadrangle $PQRS$ beccomes a parallelogram $P'Q'R'S'$, with diagonals $P'R'$ and $Q'S'$ intersecting at B'. But then $B'D'$ is a midline of the parallelogram, and since projections preserve cross ratios, we get

$$(PS, DC) = (P'S', D'C') = \frac{\overline{P'D'}/\overline{D'S'}}{\overline{P'C'}/\overline{C'S'}} = \frac{1}{-1} = -1.$$

17. The points D, E, and F are collinear and lie on the sides BC, CA, and AB, respectively, of triangle ABC. The line BE cuts CF at X, the line CF cuts AD at Y, and the line AD cuts BE at Z. Prove that AX, BY, and CZ are concurrent by projecting the line BY to infinity.

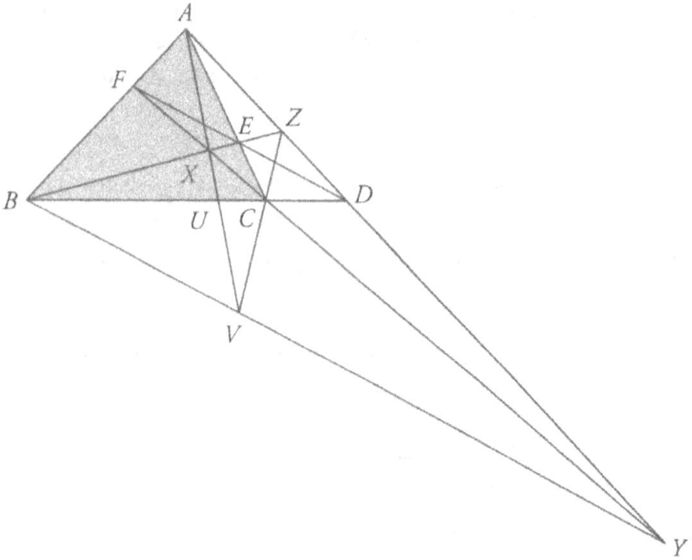

Solution. Project BY to infinity, then $ADCF$ becomes a parallelogram, and its diagonals intersect at E, as in the figure below.

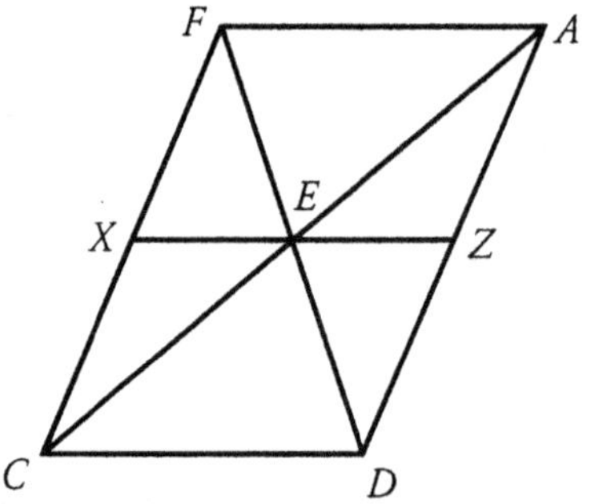

Now, XZ passes through E, and is parallel to AF. Hence, X is the midpoint of CF, while Z is the midpoint of AD.

It follows that $AZ = CX$, so that $AXCZ$ is also a parallelogram. Hence, AX and CZ are parallel, and are concurrent with BY at an ideal point V.

19. Let π be a plane that cuts an oblique circular cone in such a way that one of the generating lines of the cone is parallel to π. Prove that the intersection of π with the cone can be described by the Cartesian equation $y = kx^2$.

Solution. Let α denote the curve of intersection of the plane π with the cone. Let P be a point of α, and let ω be the circular cross-section of the cone that contains P.

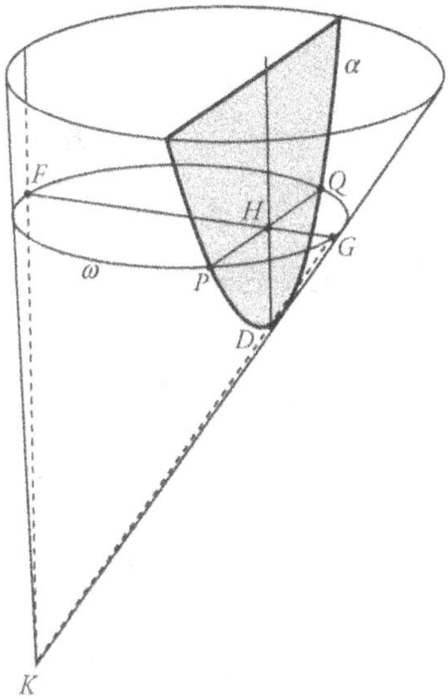

Denote the vertex of the cone by K. One of the rays generating the cone is parallel to the plane π and cuts the circle ω at F, that is KF is parallel to the plane π.

Let FG be the diameter of ω, let PQ be the chord formed by the intersection of α and ω, and let $H = PQ \cap FG$. Note that $PH = HQ$ and $PQ \perp FG$, so that by the power of the point H with respect to the circle ω we have

$$PH^2 = FH \cdot HG. \qquad (*)$$

Now let D be the point of α such that DH is parallel to the generator KF. Note that D is the same for all points P on α; that is, KD is constant.

Referring to the figure above, let $HD = y$ and $PH = x$. Using similar triangles DHG and KFG we have

$$\frac{HG}{HD} = \frac{FG}{FK} = \frac{FH}{KD},$$

so that

$$HG = \frac{FH}{KD} \cdot HD,$$

which implies

$$FH \cdot HG = \frac{FH^2}{KD} \cdot HD.$$

Hence, from $(*)$ we have

$$PH^2 = \frac{FH^2}{KD} \cdot HD$$

But FH is constant because DH is parallel to KF, and since KD is also constant, we get

$$HD = k \cdot PH^2,$$

that is,

$$y = kx^2.$$

www.ingramcontent.com/pod-product-compliance
Lightning Source LLC
LaVergne TN
LVHW051838200926
848354LV00001B/55